About Island Press

Island Press is the only nonprofit organization
in the United States whose principal purpose is
the publication of books on environmental issues
and natural resource management. We provide
solutions-oriented information to professionals,
public officials, business and community leaders,
and concerned citizens who are shaping responses
to environmental problems.

In 2006, Island Press celebrates its twenty-second
anniversary as the leading provider of timely
and practical books that take a multidisciplinary
approach to critical environmental concerns. Our
growing list of titles reflects our commitment to
bringing the best of an expanding body of literature to the environmental community throughout
North America and the world.

Support for Island Press is provided by the Agua
Fund, The Geraldine R. Dodge Foundation, Doris
Duke Charitable Foundation, The William and
Flora Hewlett Foundation, Kendeda Sustainability
Fund of the Tides Foundation, Forrest C. Lattner
Foundation, The Henry Luce Foundation, The
John D. and Catherine T. MacArthur Foundation,
The Marisla Foundation, The Andrew W. Mellon
Foundation, Gordon and Betty Moore Foundation,
The Curtis and Edith Munson Foundation, Oak
Foundation, The Overbrook Foundation, The
David and Lucile Packard Foundation, The
Winslow Foundation, and other generous donors.

The opinions expressed in this book are those
of the author(s) and do not necessarily reflect
the views of these foundations.

Resilience Thinking

Resilience Thinking

Sustaining Ecosystems and People
in a Changing World

Brian Walker & David Salt

Foreword by Walter V. Reid

⬤ **ISLAND**PRESS
WASHINGTON · COVELO · LONDON

Library of Congress Cataloging-in-Publication data.

Walker, B. H. (Brian Harrison), 1940-
   Resilience thinking : sustaining ecosystems and people in a changing world /
Brian Walker and David Salt ; foreword by Walter Reid.
      p. cm.
   Includes bibliographical references.
   ISBN 1-59726-092-4 (cloth : alk. paper) -- ISBN 1-59726-093-2 (pbk. : alk.
paper)
   1. Natural resources--Management. 2. Sustainable development. 3. Human
ecology. I. Salt, David (David Andrew) II. Title.
   HC59.15.W35 2006
   333.7--dc22

                                                      2006009300

British Cataloguing-in-Publication data available.

Printed on recycled, acid-free paper

Design by Joan Wolbier

Manufactured in the United States of America
       10   9   8   7   6

Contents

Foreword

Humanity has been spectacularly successful in modifying the planet to meet the demands of a rapidly growing population. Between 1960 and 2000, the human population doubled and economic activity increased six-fold, yet we were able to increase food production by two-and-a-half times, resulting in declining food prices and reduced hunger. The changes made to the environment to meet our growing demands have been so sweeping that cultivated ecosystems now cover more than a quarter of Earth's terrestrial surface and as much as six times more water is held in reservoirs than flows in natural river channels.

But the gains achieved by this reengineering of the planet have not been without costs, and as we enter the twenty-first century these costs are mounting rapidly. It is now widely apparent that humanity's use of the biosphere is not sustainable. Human-induced climate change is the best-known environmental threat but it is just one of a long list of challenges that already take a severe toll on people and the planet. Some 10 to 20 percent of dryland ecosystems are degraded and unable to meet the needs of people living in them; most marine fisheries are either on the verge of overharvesting or have already collapsed; billions of people face problems of water scarcity and poor water quality; and more than half of the world's ecosystem services (such as benefits ecosystems provide by purifying water, protecting coasts from storms, or helping to lessen the magnitude of floods) have been degraded.

What has gone wrong, and how can it be fixed? Without doubt, a part of the problem is that the human species is living beyond its means on a planet with finite resources: demand is out of balance with supply. Unless we use resources far more efficiently (and produce much less pollution) we will continue to erode the resource base on which our survival ultimately depends. Recognizing this problem, virtually all of our current environmental and resource management policies seek to reestablish the balance between supply and demand. We create laws and regulations such as fishing quotas to reduce use of certain fragile resources, develop new technologies such as drip irrigation to enable the more efficient use of other resources, or find new ways to manage ecosystems that enhance the production of resources such as the application of fertilizers.

These steps are all needed, but there is one catch: they won't solve the problem.

Imagine you are on a boat docked in a calm harbor and you want to quickly carry a brim-full cup of water across a stateroom without spilling. Now imagine the same situation but with the boat in rough seas. In harbor, the solution is simple: just walk quickly, but not so quickly that the water spills. At sea, speed is a secondary concern; now the real challenge is to maintain balance on an abruptly pitching floor. The solution now is to find secure handholds and footholds and to flex your knees to absorb the roll of the boat. In harbor, the solution is a simple optimization problem (walk as fast as possible but not too fast); at sea the solution requires you to enhance your ability to absorb disturbance—that is, enhance your resilience against the waves.

Since the time of the agricultural revolution, the problem of environmental management has been conceived to be an optimization problem, like the example of carrying the water on the boat in the harbor. We have assumed that we could manage individual components of an ecological system independently, find an optimal balance between supply and demand for each component, and that other attributes of the system would stay largely constant through time.

But, as we learn more about ecological and human systems, these assumptions are being shattered. Ecological systems are extremely dynamic, their behavior much more like the analogy of a boat at sea. They are constantly confronted with "surprise" events such as storms, pest outbreaks, or droughts. What is optimal for one year is unlikely to be optimal the next. And, the structure and function of the systems continually change through time (and will change even more rapidly in the future as global warming becomes an ever-stronger driver of change).

Quite simply, the basic framework underpinning our approach to environmental management has been based on false assumptions. In a world characterized by dynamic change in ecological and social systems, it is at least as important to manage systems to enhance their resilience as it is to manage the supply of specific products. In other words, we must apply "resilience thinking."

Strictly speaking, resilience thinking is not new—many traditional societies and small-scale rural farmers still give high priority to the need to manage their environment to reduce risks and buffer themselves from droughts or other surprises. Unfortunately, resilience thinking is a concept that is virtually absent from the academic and management institutions that dominate large-scale resource management practices today.

One notable exception is a group of ecologists and social scientists who began working on these issues more than fifteen years ago through a network they named the Resilience Alliance. Research undertaken through this network quickly emerged as some of the most exciting and provocative work in the biological and social sciences. The Resilience Alliance is the crucible in which an entirely new paradigm for understanding and managing our environment is taking form.

This book provides a window into this rapidly emerging field and framework. Drawing heavily on case studies and research undertaken through the Resilience Alliance, and written in nontechnical and engaging style, *Resilience Thinking* takes these issues out of their academic setting and translates them directly into practical guidance for people interested in their application in the real world. For many readers, this book will introduce concepts and ideas that will take time to fully understand. But that is to be expected from a book that is presenting an entirely new way of thinking about the management of our environment.

Resilience Thinking is, above all, a book focused on solutions. Last year saw the publication of the largest-ever assessment of the health of the planet's ecosystems—the Millennium Ecosystem Assessment. That assessment presented an extremely worrisome picture of the growing human costs associated with environmental degradation. But it also showed that it is within the power of human societies to reverse this degradation. *Resilience Thinking* provides case studies demonstrating real-world solutions to some of the most vexing environmental problems and it provides a road map for how we can solve other problems of environmental management. It is clear that the environmental problems we face will not be solved using the failed approaches of the past. But they can be solved if we embrace resilience thinking.

Walter V. Reid

Stanford University

Director of the
*Millennium Ecosystem
Assessment*

Preface

Resilience is the capacity of a system to absorb disturbance and still retain its basic function and structure. This sounds like a relatively straightforward statement but when applied to systems of humans and nature it has far-reaching consequences.

Resilience Thinking arose out of appeals from colleagues in science and industry for a plainly written account of what resilience is all about, and how a resilience approach to managing resources differs from current practices. "Don't give me complicated theory, just give me five good case studies, then I'll know what you mean" was the explicit request from one industry colleague. So that's what we set out to do.

After various iterations, reviews, and much redrafting, the "five-good-case-studies book," as it became known, has evolved into the book now before you.

As a science writer, David has worked to remove jargon and boring qualification, while injecting plain-speaking vitality into the prose. As a scientist, Brian has had the specter of colleagues peering over his shoulder and pointing at unsubstantiated statements and loose definitions. We hope we have balanced this tension appropriately because the need for a broad understanding and engagement with resilience has never been greater.

Increasingly, cracks are appearing in the capacity of our communities, ecosystems, and landscapes to provide the goods and services that sustain our well-being. Our resource base, planet Earth, is shrinking while our population continues to expand. The response from most quarters has been for "more of the same" that got us into this situation in the first place: more control, more intensification, and greater efficiency.

Resilience thinking offers a different way of understanding the world around us and of managing our natural resources. It explains why greater efficiency by itself cannot resolve our resource issues, and it offers a constructive alternative that creates options rather than limits them.

Resilience Thinking should appeal to anyone interested in dealing with risk in a complex world. This includes business leaders, policy makers, resource managers, and politicians. It also includes farmers, conservationists, community leaders, scientists, students, and members of the public—the people who elect the leaders and politicians. This is because, at its core, resilience is about risk and complexity, things that impact all of us.

We live in a complex world. Anyone with a stake in managing some aspect of that world will benefit from a richer understanding of resilience and its implications.

Acknowledgments

We thank the James S. McDonnell Foundation for its grant to Brian and Resilience Alliance colleague Steve Carpenter that enabled the development of this book. We are indebted to four reviewers Doug Cocks, Barry Gold, Lindy Hayward, and John Mitchell-Adams for their valuable feedback and also three anonymous reviewers of an early, partial draft. Their advice was most helpful. Thanks also to Yvette Salt for producing the case study maps and preparing most of the other figures.

This book is a product of the Resilience Alliance, a group of organizations and individuals involved in the interlinked aspects of ecological, social, and economic research. It is the entire group that has created and developed the framework of "resilience thinking."

There is now a substantial amount of work on resilience—hundreds of scientific papers and several textbooks, many produced by the members of the Resilience Alliance. However, until now its discussion largely has been contained within scientific circles.

The Resilience Alliance has two areas of activity—the development of the science and the communication of that science. This book is an effort toward the latter, and seeks to bring the messages and insights of resilience thinking to a wide audience.

We acknowledge the contributions of all of our Resilience Alliance colleagues but in particular Nick Abel, Steve Carpenter, Carl Folke, Lance Gunderson, Terry Hughes, Garry Peterson, Per Olsson, and Paul Ryan for their feedback and assistance in preparing the case studies. Paul in particular spent many hours with David explaining the nuances of resilience and its consequences and we owe him a large debt of thanks.

We especially want to acknowledge C. S. "Buzz" Holling, the originator of many of the ideas in this book; he is a great scientist and an inspirational mentor and leader.

Finally, we would like to acknowledge CSIRO Sustainable Ecosystems for its strong support and for hosting David while the book was being written. Gungahlin Homestead, the division's headquarters and formerly a grand farm house, was a stimulating and fitting place in which to write about the resilience of social-ecological systems and sustainability.

                                        Brian Walker and David Salt
                                        October 2005

1

Living in a Complex World:

An Introduction to Resilience Thinking

Life is full of surprises. Sometimes we take them in stride; some times they trip us up.

Consider these questions: In business, why is a competitor's new product sometimes only a minor hiccough but at other times a major shock that can destroy an enterprise? In industry, how is growth sometimes unaffected by medium interest rate rises but at other times the smallest change brings things crashing down? Why is it that the same drought that causes serious degradation of resources on one farm has little effect on another?

The response of any system to shocks and disturbances depends on its particular context, its connections across scales, and its current state. Every situation is different; things are always changing. It's a complex world.

We are all managers of systems of one type or another. That system might be a home, a company, or a nation. You might have responsibility of caring for a nature reserve, developing a mining operation, or planning fishing quotas. Be it a farm, a business, a region, or an industry, we are all part of some system of humans and nature (social-ecological systems).

How do you approach the task of management in this complex world? Do you assume things will happen in much the same way tomorrow as they did yesterday? Are you confident the system you are working in won't be disrupted by little surprises? Do you appreciate what's needed for a system to absorb unexpected disturbances?

All of these questions relate to *resilience*, the ability of a system to absorb disturbance and still retain its basic function and structure. They also relate to concepts of sustainability and the challenge of servicing current system demands without eroding the potential to meet future needs. We live in a time of growing population coupled with a declining resource base and great uncertainty about a range of environmental issues such as climate change. How can we make the systems that we depend upon resilient?

But before we address issues of resilience, stop and consider for a moment our current practices of resource management.

The Drivers of Unsustainable Development

Our world is facing a broad range of serious and growing resource issues. Human-induced soil degradation has been getting worse since the 1950s. About 85 percent of agricultural land contains areas degraded by erosion, rising salt, soil compaction, and various other factors. It has been estimated (Wood et al. 2000) that soil degradation has already reduced global agricultural productivity by around 15 percent in the last fifty years. In the last three hundred years, topsoil has been lost at a rate of 300 million tons per year; in the last fifty years it has more than doubled to 760 million tons per year.

As we move deeper into the twenty-first century we cannot afford to lose more of our resource base. The global population is now expanding by about 75 million people each year. Population growth rates are declining, but the world's population will still be expanding by almost 60 million per year in 2030. The United Nations projections put the global population at nearly 8 billion in 2025. In addition, if current water consumption patterns continue unabated, half the world's population will live in water-stressed river basins by 2025.

The Food and Agriculture Organization of the United Nations (FAO) 2004 Annual Hunger Report estimates that over 850 million people suffer from chronic hunger. Hunger kills 5 million children every year.

The most famous fisheries in the world have collapsed one after the other, including those managed with the explicit aim of being sustainable (like the cod fisheries at Grand Banks, Newfoundland in 1992). Productive rangelands are turning into unproductive expanses of woody shrubs. Half of the world's wetlands have been lost in just the last century. Lake systems and rivers everywhere are experiencing algal blooms and a raft of problems associated with the oversupply of nutrients.

The World Wide Fund for Nature's (WWF) Living Planet report

BOX 1 A Few Stats on a Shrinking World

As far as humans are concerned, Earth is shrinking. The human population is growing but the resource base required to feed, clothe, and house this growing number of people is not. Indeed, in many instances it is declining. Here are a few numbers extracted in June, 2005 from the recently released Millennium Assessment, (www.millenniumassessment.org), and from the EarthTrends website, (http://earthtrends.wri.org.), maintained by the World Resources Institute.

- Worldwide, humans have already converted nearly a third of the land area—almost 3.8 billion hectares—to agriculture and urban or built-up areas. Most of the remainder is too dry for agriculture.
- Between 1960 and 2000, the demand for ecosystem services (benefits provided by ecosystems) grew significantly as world population doubled to 6 billion and the global economy increased more than six fold. To meet this demand, food production increased by roughly 2.5 times, water use doubled, wood harvests for pulp and paper production tripled, installed hydropower capacity doubled, and timber production increased by more than half.
- Global grain production, currently 1.84 billion tons annually, will need to increase by around 40 percent to meet demand in 2020.
- The average annual growth rate of cereal production in developing countries has dropped from 2.5 to 1 percent per year over the past 35 years. Water scarcity and land degradation are already severe enough to reduce yields on about 16 percent of agricultural lands, especially cropland in Africa and Central America, and pasture in Africa.
- In the last few decades approximately 20 percent of the world's coral reefs were lost, an additional 20 percent were degraded. In the Caribbean, 80 percent of coral has been lost in recent decades. Additionally, approximately a third of the world's mangrove areas were lost.
- The number of species on the planet is declining. Over the past few hundred years, humans have increased the species extinction rate by as much as 1,000 times over background rates typical over the planet's history. (The background extinction rate is the relatively constant rate—excluding major extinction events—at which organisms have been disappearing from the fossil record over the course of geological time.)
- Since 1750, the atmospheric concentration of carbon dioxide has increased by about a third (from about 280 to 376 parts per million in 2003), primarily due to the combustion of fossil fuels and land use changes. Approximately 60 percent of that increase (60 parts per million) has taken place since 1960.
- The use of two ecosystem services—capture fisheries and freshwater— is now well beyond levels that can be sustained even at current demands, much less future ones. At least one quarter of important commercial fish stocks are overharvested. From 5 percent to possibly 25 percent of global freshwater use exceeds long-term accessible supplies and is now met either through engineered water transfers or overdraft of groundwater supplies.

analyzes the eco-footprint of 150 countries around the world every two years. In its 2004 report it estimated that the average eco-footprint around the world was 2.2 global hectares per person (a global hectare is a hectare of biologically productive space with world-average productivity). However, there are only 1.8 global hectares available per person. This ecological overshoot means we are using the equivalent of about 1.2 planets or it takes 1.2 years to regenerate what humanity uses in one year. We are using nature more rapidly than it can regenerate.

Regrettably, like a cracked record, the story goes on and on, disturbingly repetitive (see also box 1, "A Few Stats on a Shrinking World").

You've seen or heard these claims before and it is not our intention to add to doom-and-gloom publications. Rather, this book is about options and hope based on a different way of doing things through understanding how the world really works. But we do need to keep in mind what is happening to the world. The imperative message is that the world is shrinking: the human population is growing while its resource base declines.

What lies behind this decline? There is, of course, no single underlying reason; instead, there is a broad spectrum of causes. But they can be grouped into three categories: in some situations people have no choice but to overuse their resource base; in others the decline is allowed to occur willfully; and the third driver of unsustainable development is misunderstanding—the application of inappropriate models of how the world works.

The first category (no choice) relates to problems associated with large populations coupled with poverty. In this case, no other option exists than to overuse resources. It's simply a matter of survival.

All too often, however, there is a choice, and a resource is allowed to decline or is purposely driven down. Sometimes rules and regulations encourage people to overuse resources, this is the case of subsidies for drought-stricken farmers. Often these farmers are either operating on marginal land or mismanaging resources but their operation is propped up by government payments designed to protect people from hardship. In other cases, tax breaks or industry support can lead to rapid loss of a forest or a fishery. These are what are known as "perverse incentives" (McNeely 1988). Furthermore, people sometimes deliberately choose to degrade a resource because they believe science and technology will always be able to come to the rescue.

In many cases, however, resource degradation is simply the result of humankind's insatiable desire to produce and consume, leading to willful short-term greed and corruption with no heed for the future. Some suggest this is just the way humans evolved—in a world without limits where success was based on maximizing your return. Human behavior is shaped strongly by drives from our evolutionary past (competition, territory, and power) without which we would not be here as a species or as the cultures we now have. Such evolutionary antecedents made sense when the human population was small and the world was seemingly endless but this is no longer the case. In today's world such behavior has begun to turn on us and will deprive future generations of the opportunities we enjoy.

But there is a third driver as well. Our environmental problems can't all be blamed on greed and overexploitation. Ignorance and misunderstanding also play a central role in the decline of our resource base. In many instances, such as in all of the case studies in this book, it's clear that in developing a resource or a region we have not understood well enough the functioning of the ecosystems involved. The people involved were not being greedy, there was no willful destruction. Many ecosystem collapses are occurring in places where enormous resources are being invested in understanding the system and where significant effort is being made to be "sustainable."

It isn't just the amount of knowledge—details about species and ecosystems—it's also the kind of knowledge. It's the way we conceive of resource systems and people as part of them. The way we currently use and manage these systems (which we describe in the following section as "business as usual") is no longer working and yet what we hear most of the time is that the solution lies in more of the same.

This book focuses on this third driver of unsustainability. The first driver (poverty) will only be resolved when the world has addressed the other two. We return to the second driver (willful excessive consumption) in the final chapter because our best hope for dealing with it also lies in a philosophy of resilience.

Despite Our Best Intentions

Why is it that, despite the best of intentions (and in contrast to the one or two recent books telling us that "everything is okay"), many of the world's productive landscapes and best loved ecosystems are in trouble?

Current "best practice" is based on a philosophy of optimizing the delivery of particular products (goods or services). It generally seeks to maximize the production of specified components in the system (set of particular products or outcomes) by controlling certain others. Those components might be grain yields, fish catch, or timber harvest. Or, if conservation is the goal, optimization might be aimed at preserving as many species as possible in a national park or reserve. In the case of grain crops it might entail planting all the available land with a single high yielding variety and then maximizing growth with chemical fertilizers and pest control, and using large-scale cropping machinery. Production is maximized by tightly controlling each aspect of the production process.

Optimizing for particular products has characterized the early development of natural resource management, particularly in agriculture. Initially, it worked. Indeed, it resulted in enormous advances in resource productivity and human welfare. Now, however, those initial successes are bedeviled by a variety of emerging secondary and highly problematic effects on all continents and in all oceans. As Ogden Nash writes, "Progress might have been alright once, but it has gone on too long."

An optimization approach aims to get a system into some particular "optimal state," and then hold it there. That state, it is believed, will deliver maximum sustained benefit. It is sometimes recognized that the optimal state may vary under different conditions, and the approach is then to find the optimal path for the state of the system. This approach is sometimes referred to as a maximum sustainable yield or optimal sustainable yield paradigm.

To achieve this outcome, management builds models that generally assume (among other unrecognized assumptions) that changes will be incremental and linear (cause-and-effect changes). These models mostly ignore the implications of what might be happening at higher scales and frequently fail to take full account of changes at lower scales.

Optimization does not work as a best-practice model because this is not how the world works. The systems we live in and depend on are usually configured and reconfigured by extreme events, not average conditions. It takes a two-year drought, for example, to kill perennial plants in tropical savannas, and it takes extreme wet periods for new ones to be able to establish. The linkages between scales and sectors (agriculture, industry, conservation, energy, forestry, etc.) often drive changes in the particular system that is being managed. And, very importantly, while minor changes are often incremental and linear, the really significant ones are usually lurching and nonlinear—like mouse plagues in Australian wheat crops, insect pest outbreaks in forests in North America, and the sudden change from a clean, clear lake to one dominated by an algal bloom.

The Paradox of Efficiency and Optimization

"Efficiency" is a cornerstone of economics, and the very basis of environmental economics. In theory, an economy is efficient if it includes all the things that people want and value. An efficient economy, in this sense, is therefore a good thing and efficiency has become to be regarded as a laudable goal in policy and management. The paradox is that while optimization is supposedly about efficiency, because it is applied to a narrow range of values and a particular set of interests, the result is major inefficiencies in the way we generate values for societies. Being efficient, in a narrow sense, leads to elimination of redundancies—keeping only those things that are directly and immediately beneficial. We will show later that this kind of efficiency leads to drastic losses in resilience.

Optimization does not match the way our societies value things either. It promotes the simplification of values to a few quantifiable and marketable ones, such as timber production, and demotes the importance of unquantifiable and unmarketed values, such as the life support, regenerative, and cleansing services that nature provides (collectively known as "ecosystem services"). It also discounts the values placed on beauty or on the existence of species for their own sakes. Whether they realize it or not, societies depend for their existence on ecosystem services. And societies also value their ability to pass these things to future generations. Optimization, however, distorts this. It reduces time horizons to a couple of decades—the limit of the time horizon for most commercial investments. Values that do not have property rights or are publicly owned are not marketed, do not generate wealth, and gain little support, even if they involve critical ecosystem services. Often not enough people understand the criticality of the life support systems—the ozone layer and climate regulation are examples.

Though efficiency, per se, is not the problem, when it is applied to only a narrow range of values and a particular set of interests it sets the system on a trajectory that, due to its complex nature, leads inevitably to unwanted outcomes. The history of ecology, economics, and sociology is full of examples showing that the systems around us, the systems we are a part of, are much more complex than our assumptions allow for.

What it all adds up to is that there is no sustainable "optimal" state of an ecosystem, a social system, or the world. It is an illusion, a product of the way we look at and model the world. It is unattainable; in fact (as we shall see) it is counterproductive, and yet it is a widely pursued goal.

It is little wonder, then, that problems arise. And when they do, rather than question the validity of the model being applied, the response has been to attempt to exert even greater control over the system. In most cases this exacerbates the problem or leaves us with a solution that comes with too high a cost to be sustained.

In the real world, regions and businesses are interlinked systems of people and nature driven and dominated by the manner in which they respond to and interact with each other. They are complex systems, continually adapting to change. Change can be fast or slow—move at the speed of viruses multiplying or of mountains rising. It can take place on the scale of nanometers or kilometers. Change at one level can influence others, cascade down or up levels, reinvigorate, or destroy.

The ruling paradigm—that we can optimize components of a system in isolation of the rest of the system—is proving inadequate to deal with the dynamic complexity of the real world. Sustainable solutions to our growing resource problems need to look beyond a business as usual approach.

As failures mount, and as more and more people become aware of them, there is a growing dissatisfaction with the ways in which natural resources are managed. What are the important qualities of a system that need to be maintained or enhanced for a system to be sustainable? Resilience thinking is an approach (part philosophy, part pragmatism) that seeks answers to these questions.

The Key to Sustainability?

What is your version of sustainability? Is it summed up by the catch phrase "reduce, reuse, and recycle" (reduce your waste, reuse what you have, and recycle everything else)? Are you impressed by notions of ecological footprints and living within the carrying capacity of the land? Are you striving for a "factor four" improvement for the future in which we double the production from half of the input? Or maybe we should be aiming for a factor ten?

These approaches encapsulate some of the more mainstream thoughts on sustainability, and they all revolve around the notion that the key to sustainability lies in being more efficient with our resources. If we can be clever enough with the way we do things we can live within the carrying capacity of our environment.

Of course, this kind of efficiency will always be an important part of any approach to sustainability. But, by itself and of itself it is not the solution. Indeed, as we will show, by itself it has the potential to actually work against sustainability. Why? Because the more you optimize elements of a complex system of humans and nature for some specific goal, the more you diminish that system's resilience. A drive for an efficient optimal state outcome has the effect of making the total system more vulnerable to shocks and disturbances.

While that might sound counterintuitive, it is the inevitable conclusion reached by many studies investigating how social-ecological systems change over time. This book aims to explain the logic behind this seemingly perverse outcome.

By way of example of the tension between resilience and efficiency, consider the rise of the "just-in-time" approach where manufacturers dispense with big stockpiles of materials. Instead, parts and supplies are delivered to a factory at the exact moment when they are needed. The system, deemed to be efficient and optimized, yields big savings in inventory expenses but is very sensitive to shocks and has resulted in some severe industry dislocations when problems up the line with materials or staff have resulted in critical supply shortages.

The bottom line for sustainability is that any proposal for sustainable development that does not explicitly acknowledge a system's resilience is simply not going to keep delivering the goods (or services). The key to sustainability lies in enhancing the resilience of social-ecological systems, not in optimizing isolated components of the system.

The debate on sustainability has come a long way in recent decades. But if we examine it through a resilience lens, it's clear that we still have a way to go.

Embracing Change—The Heart of Resilience

At the heart of resilience thinking is a very simple notion—things change—and to ignore or resist this change is to increase our vulnerability and forego emerging opportunities. In so doing, we limit our options.

Sometimes changes are slow (like population growth); sometimes they are fast (like exchange rates, or the price of food and fuel). Humans are usually good at noticing and responding to rapid change. Unfortunately, we are not so good at responding to things that change slowly. In part this is because we don't notice them and in part it's because often there seems little we can do about them. The size of the human population is a key slow variable, for example. So too is climate change. But few people believe there is anything they can do directly to influence either.

In and of itself, change is neither bad nor good. It can have desirable or undesirable outcomes, and it frequently produces surprises.

These broad statements, when applied to interacting systems of humans and nature (social-ecological systems), take on special meanings with important consequences. Resilience thinking presents an approach to managing natural resources that embraces human and natural systems as complex systems continually adapting through cycles of change.

Most of the concepts in this book are not new. Concepts of resilience and changing ecosystems have been around for decades. However, only recently have interdisciplinary groups of scientists begun to tackle the problem in earnest. The Sante Fe Institute, for example, is one well-known group that has spawned ideas about chaos theory, network dynamics, and, latterly, robustness. Another such group is the Resilience Alliance, a collection of researchers who have pooled their insights to develop a framework for understanding change in social-ecological systems. Through the efforts of groups like these, resilience thinking may provide valuable insights to sustainability.

A Roadmap to this Book

There are many ways to present a framework for resilience thinking. We have chosen to approach it by taking three steps. The first lays down a foundation for understanding, the second outlines the core of the approach, and the third begins to explore how resilience thinking might be applied to addressing challenges in the real world.

The first step involves considering a systems perspective of how the world works:

- We are all part of linked systems of humans and nature (referred to throughout this book as social-ecological systems).
- These systems are complex adaptive systems.
- Resilience is the key to the sustainability in these systems.

A traditional command-and-control approach to managing resources usually fails to acknowledge the limits to predictability inherent in a complex adaptive system. The traditional approach also tends to place humans outside the system. Resilience thinking is systems thinking, a concept that is more fully explored in chapter 2.

The second step is to develop an understanding of the two central themes that underpin resilience thinking:

- Thresholds: Social-ecological systems can exist in more than one kind of stable state. If a system changes too much it crosses a threshold and begins behaving in a different way, with different feedbacks between its component parts and a different structure. It is said to have undergone a "regime shift." This theme of thresholds and "changing too much" is discussed in chapter 3.
- Adaptive cycles: The other central theme to a resilience approach is how social-ecological systems change over time—systems dynamics. Social-ecological systems are always changing. A useful way to think about this is to conceive of the system moving through four phases: rapid growth, conservation, release, and reorganization—usually, but not always, in that sequence. This is known as the adaptive cycle and these cycles operate over many different scales of time and space. The manner in which they are linked across scales is crucially important for the dynamics of the whole set. These ideas are explored in chapter 4.

The third step is to apply this understanding to the real world:

- How might a resilience approach be put into operation?
- What are the costs of a resilience approach?
- What are the implications for policy and management?
- What might a resilient world be like?

While a framework for resilience thinking provides valuable insights into why and how systems behave as they do, to have policy and management relevance it needs to be able to solve problems in resource management, which is discussed in chapters 5 and 6. In chapter 6 we also discuss how managing for resilience has the capacity to create space in a shrinking world by opening up options rather than closing them down. Resilient social-ecological systems have the capacity to change as the world changes while still maintaining their functionality. Resilient systems are more open to multiple uses while being more forgiving of management mistakes.

While every effort has been made to keep jargon and terminology to a minimum, resilience thinking does contain several concepts that can seem a bit daunting to the uninitiated on first exposure. We encourage readers not to be too worried about understanding every detail on the first reading. Instead, try to take away a general appreciation of what thresholds and adaptive cycles are, while attempting to understand them in relation to the system in which you are interested.

Even if the finer details of some aspects of the resilience approach remain a bit obscure, if you can incorporate the broader themes presented here on living within complex adaptive systems you'll discover you've acquired a powerful set of insights about how the world works. Concepts of sustainability, efficiency, and optimization all begin to take on a new light.

Our hope is that readers will start asking questions about the systems in which they live or in which they are interested: What are the key variables driving them? Is the system approaching a threshold? What management actions do you need to consider in order to avoid such a threshold? What are the dynamics of this system? What are the connections between the scale at which you are concerned and the next level up and down?

These are all big questions that may not be easy to answer. However, the very act of framing them in relation to the system in which you play a role is an important step toward resilience thinking.

Between each chapter a case study on a region illustrates the significance of resilience thinking when applied to real-world situations. They demonstrate its value in interpreting and understanding what lies behind changes being observed in five very different social-ecological systems around the world.

Five Regions, Five Stories

The five regions we discuss are:

- The Everglades in Florida, United States: Possibly the world's most famous marshland system. Significant parts of the national park have crossed a threshold into a new regime dominated by cattails.
- The Goulburn-Broken Catchment: One of Australia's most agri-culturally productive regions. Saline groundwater now lies just beneath the surface of the region's most productive agricultural zone.
- The coral reefs of the Caribbean: Once one of the most magnificent coral systems in the world and a tourist draw that was the economic lifeblood of the region. In the last thirty years, 80 percent of hard coral reefs have disappeared and the remaining reefs are at risk.
- The Northern Highland Lakes District of Wisconsin, United States: A fishing paradise with an uncertain future. The natural amenity of this much-loved area is slowly being lost as its population grows.
- The Kristianstad Water Vattenrike: An internationally renowned wetland in southern Sweden. Its beloved wet meadows are being lost, water quality is in decline, and wildlife habitat is disappearing.

Why these five regions? To begin with, they are different. They have very little in common, with different levels and types of population engaged in different enterprises coping with a range of different challenges. What they do have in common is that each is confronted with a range of natural resource and social challenges that have major implications for their inhabitants and surrounding regions. And we know quite a lot about them; each has been studied over many years in an attempt to understand the ecological and sociological processes that drive them.

We have chosen case studies at a regional scale because this is the focus of much of the work of the Resilience Alliance. However, as the basis of resilience thinking becomes clear, it should be apparent that it applies to systems of people and nature at all levels: individuals, communities, businesses, and nations.

Of course, there are many other regions around the world facing enormous resource issues that are not discussed in this book. Many parts of Africa, for example, are suffering chronic food shortages, disease outbreaks, and social instability. Two such regions (in Mozambique and Zimbabwe) are part of the Resilience Alliance's set of case studies and there are many lessons in resilience thinking that are relevant to these regions. However to meet the needs of this book as an introduction to resilience thinking we have chosen to examine five regions that are well studied and that reflect a range of contrasting issues.

Our first case study is the Everglades, a world-renowned wildlife wonderland at the southern tip of Florida in the United States. Attempts to tame parts of it for agriculture and urban settlement over the last hundred years have had mixed results. On the one hand the region supports a lot more people, industry, and agriculture. On the other, its natural qualities have gone into steep decline, including its water quality. Development has resulted in some significant gains but the costs are only now being understood.

Key Points on Resilience Thinking

- Current approaches to sustainable natural resource management are failing us. They are too often modeled on average conditions and expectations of incremental growth, ignore major disturbances, and seek to optimize some components of a system in isolation of others. This approach fails to acknowledge how the world actually works.
- Business as usual is about increasing efficiency and optimizing performance of the parts of social-ecological systems that deliver defined benefits, but fails to acknowledge secondary effects and feedbacks that cause changes (sometimes irreversible changes) in the bigger system, including changes to unrecognized benefits. While increasing efficiency is important for economic viability, when undertaken without considering the broader system's response it will not lead to sustainability; it can lead to economic collapse.
- Resilience thinking is about understanding and engaging with a changing world. By understanding how and why the system as a whole is changing, we are better placed to build a capacity to work with change, as opposed to being a victim of it.

CASE STUDY 1

Carving up a National Icon:
The Florida Everglades

One hundred years of command and control management have exacted a heavy price on the Everglades, possibly the most recognized wetland in the world. Efforts to develop it have involved compartmentalizing it into agricultural, urban, and conservation sections; draining it; and constructing massive infrastructure to control floods and mitigate the damage from hurricanes. The results have significantly reduced the area of natural habitat, created dramatic declines in water quality (for wildlife and humans), and made the region increasingly vulnerable to the shocks produced by extreme weather events.

The Everglades of today is propped up by injections of billions of dollars from the federal government, while being held in gridlock by litigation and a highly adversarial contest between a myriad of players. It is a social-ecological system with a major resilience problem (Gunderson et al. 2002).

For all this, the Everglades is still regarded as an international icon for natural beauty. And yet the very aspects of this region that make it world-renowned are under a serious and growing threat as human development in and around the Everglades has slowly shifted the pattern of dynamics that has crafted the region. And what's at risk is not just the "nature" portions of the system in the national park, but the hydrological changes impact on the economic prosperity and social stability of the broader region that now supports over 6 million people.

The irony is that the very developments that opened up the territory for growth and prosperity, specifically the control of water levels, have made the system increasingly vulnerable to shocks. And, if the past century is anything to go by, this is an area where there's always a new surprise just around the corner. Will these World Heritage 'glades remain the "Ever" glades, or will their future be the "Never-again" glades?

The Everglades in a Nutshell

"Here are no lofty peaks seeking the sky, no mighty glaciers or rushing streams wearing away the uplifted land. Here is land, tranquil in its quiet beauty, serving not as the source of water, but as the last receiver of it. To its natural abundance we owe the spectacular plant and animal life that distinguishes this place from all others in our country." With these words, President Harry S. Truman formally dedicated the Everglades National Park in 1947. The event was the culmination of years of effort by a dedicated group of conservationists to make a national park in the Florida Everglades a reality.

The Everglades historically covered some 10,500 square kilometers of the southern tip of the Florida peninsula. The region extends two hundred kilometers along a north-south axis and is about eighty kilometers at its widest point. The Big Cypress Swamp forms its western boundary while

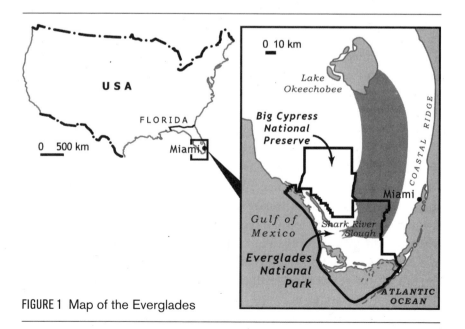

FIGURE 1 Map of the Everglades

to the east it is bordered by the raised Atlantic Coastal Ridge. Lake Okee-chobee forms the northern edge and is a crucial source of water for the Everglades ecosystem. The southern terminus consists of tidal mangrove forests along the tip of the Florida Peninsula. Here, waters from the Everglades mix with the salt waters of Florida Bay and the Gulf of Mexico.

The Everglades is a patchwork of landscapes: lakes and rivers, freshwater marshes, tree islands, mangrove swamps, pinelands, and coastal waters. Through it all flows water—lots of it, in varying amounts. The flow begins in the north with headwaters in the Kissimmee River. Rain drenches the system every summer, with an annual average of around 1.3 meters of water. This water flows into Lake Okeechobee, which is wide and shallow, covering close to 1900 square kilometers, and averaging just 3.6 meters in depth.

The rainy season historically flooded the lake's southern shores. These flood waters then moved south forming the famous "river of grass," a slow moving mass of shallow water flowing towards the ocean through marshes forming a channel up to 80 kilometers wide and 160 kilometers long.

The Everglades, therefore, is an ecosystem that has literally been fashioned and shaped by change and flowing water: by a yearly cycle of summer flooding, frequently of cyclonic proportions, followed by a protracted dry season. The plants and animals, many of them unique to the Everglades, are adapted to alternating wet and dry seasons. During the dry season (December to April), water levels gradually drop. Fish migrate to deeper pools. Birds, alligators, and other predators concentrate around the pools to feed on a varied menu of fish, amphibians, and reptiles. This abundant food source is vital to many wading birds which nest during the dry season.

In late May, spring thunderstorms signal the beginning of the wet season. A winter landscape dotted with pools of water yields to a summer landscape almost completely covered with water. Wildlife disperses throughout the region. Insects, fish, and alligators repopulate the 'glades, thus replenishing the food chain. By December, the rains cease and the dry cycle begins again.

Europeans came into this ecosystem and decided the best way to exploit its organic soils was to control the water that shaped it. With hindsight, it's little wonder that this approach was going to have profound impacts on the functioning of the whole system.

Rich as Dung Itself

While the development of the Everglades really got underway in the early twentieth century, notions of harnessing this wildlife-rich wetland had been brewing since 1769 when the English writer William Stork called the soil of the lower part of Florida "as rich as dung itself."

In 1823, a cartographer referred to the south Florida wetland as the "Ever-glades," a "glade"[*] being a term for a bright streak or a patch of light. Some believe this refers to the interplay of sunlight with the grass and the slowly moving water below. This creates a light space that marks one of the special attributes of the region. To those early European eyes the region must have appeared as a vast, forbidding, alien paradise stretching from horizon to horizon. The description of "ever" in its title suggests its vastness and watery nature had been there forever.

Of course, as we now know, the Everglades is actually a very recent formation, geologically speaking. Ten thousand years ago it was dry forest. With rises in sea level it took on the characteristics of tidal marsh, mangrove swamp, fen, lake, and finally scattered uplands with hardwood trees. Rich detritus falling through the water built up over the centuries to form the Everglades muck land—one of the largest bodies of highly productive organic soil in the world. (Though, as was to be discovered, the system as a whole was nutrient poor with low levels of available phosphorus being a major limiting factor that configured the system.)

Draining the Swamp

The potential of the region for agriculture wasn't lost on the American generals and soldiers who had over many years pushed the local Indians deeper and deeper into the south Florida interior during the first half of the nineteenth century. In 1848, the first proposal to drain the Everglades was submitted to the U.S. Senate, a proposal that was both welcomed and encouraged. However, the first real efforts to drain the swamps didn't take place till the late years of the nineteenth century.

In the early 1900s, the Everglades were viewed as a big, mean swamp. As author James Carlos Blake described it, "If the Devil ever raised a garden, the Everglades was it." That view was seemingly borne out by

---

[*]Another interpretation of "glade" is that it is a clearing or patch in the forest, hence Everglades may also refer to a large clearing. Native Americans referred to the area as a "grassy lake."

IMAGE 1

A Landsat image of the Everglades National Park showing the rich mosaic of water-ways, saw grass prairie, mangrove and cypress swamps, pinelands, and hardwood hammocks. The effects of highways on the water regime are very apparent.

Photo credit: Image courtesy of NASA Landsat Project Science Office and USGS EROS Data Center.

the flood of 1903, which destroyed most of the region's farms. In 1906, Florida Governor Napoleon Bonaparte Broward began to make good on a campaign promise to create an "Empire of the Everglades," by wringing the last drop of water out of that "abominable, pestilence-ridden swamp." He initiated the first engineering campaigns to reroute the wetlands' waters to the Atlantic, and dry the land out for farming.

In 1908, drainage operations began opening up new tracts of land for cultivation and the future looked bright. Partial drainage of the Everglades spurred dramatic growth in South Florida; newcomers were lured by cheap land and favorable reports by northern newsmen touting the region's glamour and financial promise. Numerous small towns arose, particularly around the shores of Lake Okeechobee. To protect them from flooding a small levee along the southern shore of the lake was constructed. The structure, composed largely of muck and sand, stood a couple of meters high and ran some seventy-five kilometers along the lake's edge.

Gone with the Wind

And then in 1926, as had happened every couple of decades, a big hurricane scored a direct hit on the region. With winds in excess of two hundred kilometers per hour, storm waters spilled over the levee wall destroying thirteen thousand homes and farms and killing over four hundred people.

There were several major storm events over the following couple of years but it was the next big one, which struck in 1928, that really shaped the future of this area. Winds of 215 kilometers per hour whipped up a wall of water from Lake Okeechobee that drowned nearly two thousand people in one hour making it one of the worst storms in the history of the United States. It brought home the crushing realization that draining the swamps was only a partial solution to taming this region and exploiting its natural wealth. If agriculture was to flourish it needed to be protected from these extreme events.

With funding and support from the federal government, including the personal endorsement of President Herbert Hoover, the Army Corps of Engineers responded with a major flood control project, constructing levees that bounded Lake Okeechobee like a Berlin Wall. The Hoover Dike, as it was called, flanked three quarters of Lake Okeechobee, and was completed in 1938 at a cost exceeding $23 million. The top of the dike extended some six meters above the normal lake level. Reassured by bricks and mortar, agriculture reestablished itself with renewed vigor.

The 1940s, however, once again demonstrated the region's vulnerability to extreme events. Severe drought, following decades of drainage, led to dramatic wildfires across the Everglades in 1943 and 1944. Large quantities of the region's precious organic soil were consumed by the fires and lost forever.

In 1947 and 1948 tropical storms and hurricanes again struck southern Florida. This time more than two thousand people and twenty-five thousand cows were drowned. An incredible 2.7 meters of rain fell on the region in six months, submerging farm, field, and town.

The crisis gave rise to a belief that the solution to flooding was even more control of the water, and gave birth to a new era of management. Under a large bureaucratic plan called the Central and Southern Florida Project for Flood Control and other Purposes (C&SF), the Army Corps of Engineers erected levees, canals, and pumps capable of controlling 13.8 billion liters of water per day.

By 1963, the C&SF put in place operational procedures and created dedicated land use areas: the Everglades Agricultural Area, Water Control Areas, and the Everglades National Park. It also sparked a surge in South Florida's human population. The C&SF reflected a partnership of federal, state, and local governments in control of the hydrologic infrastructure. It became more and more interconnected, building in a rigidity that, as events would demonstrate, primed the Everglades for further disaster.

And Then There Was Drought

Surprise came in the 1960s and 1970s, with the arrival of droughts—not the planned-for floods. The dried-out watershed erupted once again in flames. Through the heat and smoke came the realization that decades of management simply weren't working, and the natural system that lay at the foundation of the area's prosperity was ailing.

A new era began in 1971, with attempts to put in place mechanisms to deal with water shortages as well as floods. Rather than creating physical capital, the emphasis during this cycle was on creating social capacity through an institutional reconfiguration that formed the South Florida Water Management District. There was now heightened awareness of water supply and environmental problems in south Florida. The Governor's Conference on Water Management in south Florida concluded that water quality was deteriorating significantly and that water quantity was insufficient during the dry season.

The 1970s was also a time of big declarations on setting aside land and resources for nature reserves. In 1971, Congress set the minimum water flow to Everglades National Park at 390 million cubic meters per year following several years of extreme dry conditions. In 1974, the Big Cypress National Preserve was established. In 1976, UNESCO recognized the Everglades National Park as part of the international network of biosphere reserves. In 1979 the Everglades was designated as a World Heritage Site.

Paradoxically, it was also a time of declining biodiversity. In the late 1960s several iconic species were listed as endangered including the Florida panther, the snail kite, and the Cape Sable seaside sparrow. The American crocodile, whose entire North American range is in and around Florida Bay and Biscayne Bay, was placed on the federal endangered-species list in 1975. At the time, the population was listed as two hundred animals, with just ten nesting females.

In the 1970s it was also noticed that an uncommon plant in this region, known as cattail or bulrush, was taking over wide areas of marshland that were previously dominated by sawgrass. This, in turn, led to the decline of many species of wading birds that used the sawgrass for nesting. The natural values of the Everglades were under siege on a number of fronts.

A closer examination of what was producing this transition, however, revealed that the marshland ecosystem was responding to a number of cycles operating over different scales, and that the invasion of the cattails was a legacy of changes that had been instigated some thirty years earlier. The saga of the cattails is also a lesson in what happens when ecological resilience declines and thresholds are crossed.

Shifting Marsh Grasses

The threshold involving the cattails involves nutrient enrichment (Gunderson 2001). Historically, the Everglades was a low nutrient wetland. Low levels of phosphorus in the water were a major limiting factor for plant growth. For the past five thousand years, the ecosystem effectively self-organized around this low level of nutrients. Nutrients were often released into the system following fires that usually occurred when the marshes experienced several dry years in a row. This occurred on average once every ten years.

The combination of this low level of nutrients, decadal fire cycle, and annual wet/dry cycle led to the formation of a landscape mosaic with small areas of enhanced nutrients in tree islands (the nutrients being maintained by the nesting of wading birds). The plants in the remainder of the Everglades landscape (sawgrass marshes and wet prairies) were adapted to low levels of nutrients.

When the Everglades was split into three designated land uses (agriculture in the northern third; urban usage in the eastern fifth; and conservation in the southern and central remaining half), the latent impacts of this compartmentalization didn't become visible until the late 1970s and early 1980s. During this period, large-scale shifts occurred in the vegetation from a sawgrass- to cattail-dominated marsh in the areas immediately south of the agricultural area.

The transition was attributed to a two-part process. First, through fertilizer-enriched water flowing out of the northern agricultural part, there was a slow increase in the concentration of soil phosphorus levels. This was in part due to government-provided price support to U.S. sugar that drove the spread of sugarcane in Florida's agricultural area. In the 1960s and '70s farmers had the political power to move water into Lake Okeechobee during floods. Eutrophication in the late 1970s led to the delivery of phosphorous-rich water to the wetlands of the Everglades.

This was then followed by a fire, drought, or freeze that killed some of the sawgrass and opened up the opportunity for its replacement by cattail. Cattail cannot survive under low phosphorus, but under high phosphorus, if it can get a foothold, it will easily out compete sawgrass.

The change reveals the complex nature of how an ecosystem shifts over time, and how these shifts are driven by processes that work at different scales in time and space. The vegetation structures represent the most rapidly changing variables, with plant turnover times on the order of five to ten years.

Fires, droughts, and freezes have all been part of the configuring processes (or disturbances) around which the Everglades has self-organized for over fifty centuries. The disturbance regimes of fires operate at return frequencies between ten to twenty years. Other disturbances such as freezes and droughts occur less frequently at return times of several decades.

The soil phosphorus concentrations are the slowest of the variables, with turnover times on the order of centuries. The resilience of the freshwater marshes is related to the soil phosphorus level.

The original composition of sawgrass and wet prairies was maintained by the interaction of fire, droughts, and the annual water cycle. When hit by severe fires, peat was combusted and removed. This lowered the level of the soil. Wet prairie communities would then replace sawgrass communities on the lower wetter sites. Without recurrent, severe fires, the sawgrass community would slowly colonize the wet prairie community.

The composition of this shifting mix had little or no cattails, except around alligator holes, or other areas where animals concentrated nutrients. It is only since the late 1970s that cattail has formed extensive single-species stands in areas of nutrient enrichment, covering approximately 3 percent of the remnant Everglades.

Cattails colonized sites where disturbances had removed or top-killed sawgrass plants and where soil phosphorus concentrations were greater than thirty parts per million. Rapid regeneration by easily dispersed seeds and faster growth rates of cattails give it a competitive advantage over the sawgrass.

Up until the 1970s the sawgrass/wet prairie marshlands were resilient to the disturbances of fires, droughts, and freezes (i.e., these systems could absorb these types of disturbances and still retain their basic function and structure). From the 1970s on, however, they had lost this resilience, and the ecosystem moved into a new regime, one that came as a considerable ecological surprise.

A Sense of Crisis

Droughts and flooding continued in the early 1980s, until the problems were so severe—and the public outcry so great—that Governor Bob Graham initiated a Save Our Rivers campaign. The campaign recognized that the entire ecosystem needed to be restored, not just parts of it. Over the following years, laws were enacted and agencies given powers to protect and rehabilitate Florida's lakes, bays, estuaries, and rivers. As the sense of crisis grew so too did the adversarial nature of the interaction between the different levels of government. In 1988, the federal government filed a lawsuit against the state of Florida, the South Florida Management District, and other state agencies for polluting the Everglades with excessive phosphorus.

In the following year, President Bush, Sr., signed into law the Everglades National Park Protection and Expansion Act of 1989, authorizing the addition of four hundred square kilometers of the east Everglades to the park. The act also sought to restore the natural hydrological conditions within the park.

The early 1990s saw a variety of proposals and laws put forward to restore and preserve south Florida's ecosystem, enhance water supply, and maintain flood protection. In 2000, the Comprehensive Everglades

Restoration Plan was passed. It includes dozens of projects estimated to take over thirty years to complete and comes with a price tag of almost $8 billion (to be shared by the federal and state governments), a staggering amount of money.

Unlike previous schemes, the current interventions are influenced by a better understanding of how the ecosystem functions. In a significant about face, rather than add further layers of control infrastructure, the latest plans call for the removal of some of the artificial structures that obstruct natural water flow. Around four hundred kilometers of levees and canals will be removed, while thirty-two kilometers of the Tamiami Trail, the major east-west highway across Florida, will be raised off the ground to allow the river to flow freely under it. It marks the first time the U.S. government has reversed a major public works program out of consideration for the environment (and a recognition that the early command and control approach was not working).

And yet the same plan is also proposing many grand engineering solutions. The boldest idea calls for the capture of about 6.5 billion liters of fresh water now expelled every day through the canals into the Atlantic Ocean and Gulf of Mexico. The water would be injected into Florida's aquifer 335 meters below the surface, where it will be stored and then released when needed into the Everglades. This concept has never been tested on such a large scale.

The Wash Up

One hundred years of a command and control development has reduced the Everglades in size by 50 percent, reduced flows to the remaining wildlife areas by 70 percent, and grossly reduced water quality. Today, sixty-eight species in the Everglades are endangered and portions of the marshes have been lost, possibly forever, to cattail. Exotic weeds such as kudzu vine and callitris pine are also taking over significant areas of the park. The only thing propping up the region is massive injections of money from a higher level of government.

Agriculture and the human population continue to grow, and along with it, the need for water. The population has grown from 10,000 to more than 6 million over the past century and some projections suggest this might double to 12 million by 2050.

The development of the Everglades has been characterized by a series of surprises and resource crises, and these crises reveal a clear failure in both understanding and resource policy. Looking at the sequence of events over the past hundred years it becomes apparent the social-ecological system that is the Everglades has been going through cycles of incremental growth, crisis and reformation. Every so often an existing approach to development is thrown into disarray by a crisis. Each major crisis has precipitated a release of resources, the unraveling of existing policies, and the development of a new set of policies. It is following these periods of crisis that institutions and the connections between them are most open to dramatic transformation.

Crisis and disturbance have shaped and driven the Everglades, biophysically and socially. Indeed, the Everglades of today is mired in a web of litigation as conservationists, farmers, industry groups, and government agencies at many levels fight it out in the courts to see whose rights are being infringed and who is to be held liable. It is a legal arena that entices aspiring young lawyers (to paraphrase the title of a famous stage play: "Oh, what a lovely war this is!"). In such an adversarial climate it's extremely difficult to introduce wide-ranging innovation and reform. Maybe it will take the next big one (flood, drought, or fire) to create the space in which things might happen.

Mention the Everglades anywhere and people think of a magnificent marshland of international significance. Look a bit closer, however, and it quickly becomes apparent that to understand and effectively manage and protect its many resources you need to see it as a complex social-ecological system operating over many linked cycles.

The next chapter spells out what this means and those that follow lay out what a resilience approach would advocate. For the Everglades it involves trying to identify the regime shifts that are possible, the feedback changes that determine the thresholds between regimes, and the attributes of the system that govern the controlling variables and strengths of the feedbacks. What all this means will become clearer as the book unfolds.

Resilience and the Everglades

As is often the case when attempting to understand what makes an ecosystem resilient, the real impacts of draining and compartmentalizing the Everglades didn't become apparent for decades. However, when cattails began displacing sawgrass it did become clear that the system was behaving in a different way. The usual set of disturbances of drought, fire, or freeze would not normally have seen the loss of sawgrass but after the 1970s this is what happened. Closer examination revealed that the controlling variable was the amount of phosphorus in the soil, and the increase in phosphorus was due to changes in land use from decades earlier.

Appropriate development of the Everglades demands an understanding of how it functions as a system; it requires seeing it as a linked social-ecological system. Just as the ecological domain is shaped by a range of linked cycles operating over different scales of time and space, so too is the social domain. It also consists of linked cycles of growth, crisis, and reorganization.

2

The System Rules:

Creating a Mind Space for

Resilience Thinking

Humans are great optimizers. We look at everything around us, whether a cow, a house, or a share portfolio and ask ourselves how we can manage it to get the best return. Our *modus operandi* is to break the thing we're managing down into its component parts and understand how each part functions and what inputs will yield the greatest outputs.

Over the short term it's not a bad way to go. Over the longer term, and thinking of the impacts on the broader system, a narrow focus on optimizing for some particular product or thing creates a raft of significant problems.

Consider the Everglades case study. Developers saw it as a rich muck land in need of a bit of regulation. They saw big returns; all that was needed was some drainage and flood infrastructure. In the short term they were right. Over the longer term it became obvious that the region's character was being written by extreme events of flooding and drought. To maintain the investment required a greater and greater input of resources, and with each attempt at greater control new problems were created.

What's more, the changes that had fashioned the Everglades over thousands of years were also the forces that vitalized the ecosystem. Regulating the system to optimize its short-term human returns was shutting down natural cycles that sustained its wealth and allowed it to recover from extreme events.

Who's in Command, Who's in Control?

The way we commonly do business is sometimes referred to as "command and control" as it involves controlling or commanding aspects of a system to derive an optimized return. The belief is that it's possible to hold a system in a "sustainable optimal state." It's easy to see why the term is used in cases like the Everglades in which enormous federal resources were employed to regulate and control its mighty water flows. Mountains of concrete give the impression that humans are in control.

This is not, however, how the world actually works. Yes, we can regulate portions of a system—and increase that portion's return over the short run—but we can't do this in isolation from the rest of the system. When some part is held constant, the system adapts around our changes, and frequently loses resilience in the process. While we can hold parts of the system in a certain condition, the broader system is beyond our command. Indeed, no one is in control; this is a key aspect of complex adaptive systems, and it will be discussed in more depth later in this chapter.

Governments like to think they are in control. People respond to their policies and rules (often doing their best to get around them). But governments, even dictatorships—those that survive—respond to what people want. It's summed up nicely in the title of a book on agricultural development (Godden 1997): "Governing the farm and farming the government." Who's controlling who?

Our Changing Perspective of Change

While science and technology have an enormous impact on how we view ourselves and the world, accepting change has never been smooth or easy. Throughout history, it has been easier to deny or ignore information at odds with the prevailing worldview than change to change it.

Centuries ago we considered the world flat because that is what our eyes told us. Even when boats setting off in one direction and returning from the other direction suggested that the world was round, we were loathe to give up the flat-Earth view.

When Galileo challenged the Earth-centric model of the Universe citing irrefutable evidence from his newly fashioned telescope, the Church forced him to recant on penalty of death rather than change antiquated notions.

When Darwin challenged the human-centric model of existence with the theory of evolution he ignited a debate that raged for many years (and still rages in some parts of the world).

Adopting new ways of thinking is never easy. The science philosopher Thomas Kuhn probably sums it up best with his "structure of scientific revolutions." The weight of evidence in favor of the "new" paradigm has to be overwhelming, and often the revolution is triggered by some event—like the death of a dominant proponent of the old paradigm (a view that closely reflects several aspects of adaptive cycles, one of the building blocks of resilience thinking, see chapter 4.)

Just as the notions of a sun-centric solar system and evolution were difficult to accept, so too is the notion that the living world around us is in a constant state of flux. We know that life has changed dramatically over geological timeframes because we have a rich fossil record to prove it. However, for most of us that is more of an academic notion than a reality because we don't experience time on such scales.

Day-to-day, the landscapes around us appear constant, changing incrementally if at all. Or, if some event such as a fire or a storm strikes, elements of the landscape might be damaged or disappear but they will inevitably reappear at some point (or so we think). Our basic mental model is that the ecological system around us can assimilate change unless we completely destroy it or convert it into a city. If management shifts the equilibrium considerably (e.g., regulation of water in the Everglades impacting the natural marsh balance), the system will stay in this new state and human intervention can manage the situation to ensure the new equilibrium isn't causing us too much trouble.

Indeed, over the last century some of our finest ecological minds explained through their work the mechanics of many aspects of ecosystem behavior and made real the concept of an equilibrium state: push the system one way and the system compensates and moves toward a natural equilibrium. Leave the system alone and it will eventually gravitate to some equilibrium condition. On a longer time scale, the equilibrium itself might change. Open ponds might eventually end up grassy wetlands, and open plains might change to being forests.

In many ways, this notion of ecological systems fluctuating around an equilibrium state is a powerful and enabling model. Humans living within an ecosystem can extract goods (like agricultural produce) and services (e.g., water purification) from an ecosystem just as long as they don't actually destroy it; the system will recover. If problems arise, they can be addressed by reducing or removing the human intervention that created them.

This paradigm suggests it is possible to operate in an optimizing fashion. Understand a component of the system then control it to maximize the particular output you're interested in. Over time, monitor the process and tweak your management to get the system to deliver the optimal return (in terms of costs and benefits).

However, this view of optimization is proving anything but sustainable.

Resilience thinking is an alternative way to understand social-ecological systems. It directly challenges entrenched ways of interpreting the world around us; ways that served the pioneers, but that are failing their successors. Consequently it's unlikely to be taken up without a struggle.

A System's Mind Space

As we said at the beginning of this chapter, humans are great short-term optimizers. But we're not so good over longer timeframes. That requires systems thinking. Resilience thinking is *systems thinking*. Three concepts are imperative to this kind of thinking. Within this "mind space" a resilience framework can take form and make sense.

- Concept 1: We all live and operate in social systems that are inextricably linked with the ecological systems in which they are embedded; we exist *within* social-ecological systems. Whether in Manhattan or Baghdad, people depend on ecosystems somewhere for their continued existence. Changes in one domain of the system, social or ecological, inevitably have impacts on the other domain. It is not possible to meaningfully understand the dynamics of one of the domains in isolation from the other.
- Concept 2: Social-ecological systems are complex adaptive systems. They do not change in a predictable, linear, incremental fashion. They have the potential to exist in more than one kind of regime (sometimes referred to as "alternate stable states") in which their function, structure, and feedbacks are different. Shocks and disturbances to these systems (e.g., fires, floods, wars, market changes) can drive them across a threshold into a different regime, frequently with unwelcome surprises (such as

a lake suddenly going from a state of clear water to a persistent
state of murky water).

- Concept 3: Resilience is the capacity of a system to absorb
disturbance; to undergo change and still retain essentially the
same function, structure, and feedbacks. In other words, it's the
capacity to undergo some change without crossing a threshold
to a different system regime—a system with a different identity.
A resilient social-ecological system in a "desirable" state (such
as a productive agricultural region or industrial region) has a
greater capacity to continue providing us with the goods and
services that support our quality of life while being subjected to
a variety of shocks.

Concept 1: We Are All Part of the System

We exist in linked social and ecological systems. This should be a self-
evident truth; however, it's not reflected in the manner in which we tra-
ditionally analyze and practice natural resource management. We have
economists who model "the economy," sociologists who explain how
and why human communities behave as they do, and scientists who
attempt to unravel the biophysical nature of ecosystems. They all generate powerful insights into how the world works; but these insights are
partial. They are only on components of the system rather than the system as a whole. While there have been growing calls for greater inte-
gration of these various disciplines for years, it is only just starting to
happen (Holling and Meffe 1996).

What's more, scientists, be they social, economic or biophysical,
tend to study the system from a perspective of being outside whereas
in fact they, too, are part of the system. Resilience thinking is all
about seeing the system—the social-ecological system that we're all
a part of—as one interlinked system. We are all actors playing a role
in that system.

If we accept this premise it changes the way we look at the world.
For example, a suburb and a nearby wetland are both part of the same
system. The government agency that is managing the wetland, the
developers who would like to drain it for new land development, and
the scientists who believe it provides valuable ecosystem services are
all part of the system itself, they are not outside it. Change the system in one place (e.g., drain the wetland, or change the legislation

---

BOX 2 People and Crayfish in West Sweden

Studies of a crayfish fishery in Lake Racken in west Sweden suggest that understanding and empowering the people harvesting the crayfish is just as important as understanding the ecology of the crayfish (Olsson and Folke 2001).

The noble crayfish population of the lake had decreased dramatically since the 1960s, under the multiple threats of acidification, fungal disease, and overexploitation. In 1986, local landowners formed the Lake Racken Fishing Association to address the problem, and began to implement a range of preventative measures to enhance crayfish populations by enhancing habitats and reducing threats. They limed parts of the lake, exchanged crayfish between different parts of the lake to prevent inbreeding, and imposed temporary fishing bans. Members monitored pH, calcium levels, metal concentrations, and several indicator species such as insects, mollusks, and fish. The ecosystem management applied by the association's members combined scientific knowledge and local observation, and they learned about their ecosystem as they tried different strategies. This is known as *adaptive management* or *adaptive learning*.

In response to this diversity of strategies implemented by the locals, the noble crayfish population slowly recovered but not to the levels experienced in the 1950s and 60s. For some this recovery simply isn't fast enough and alternatives have been proposed such as building crayfish hatcheries or stocking the lake with the American crayfish. These could lead to a greater initial increase in crayfish populations. Unfortunately, they require money and could lead to an erosion of the local institutions already devoted to sustaining the functioning of the lake.

Promoting the involvement of local people and supporting adaptive learning are two important aspects of managing the resilience of social-ecological systems. Adaptive learning increases local knowledge and coordination, and keeps locals involved with their lake system. In the long term, this will lead to greater success in addressing current problems and generating a greater capacity to respond to the future.

---

that protects it, or change the attitudes of the developers or change the information flows from the scientists) and you will inevitably see feedback responses elsewhere in this social-ecological system. Our current approach does not take all these feedbacks into account. We operate as if a change to the suburb, or to the wetland or to our attitudes toward the wetland can occur in isolation to the rest of the system.

Take a good look at the systems of which we are all a part and it soon becomes apparent that the biophysical system constrains and shapes people and their communities, just as people shape the bio-physical system.

Concept 2: Appreciating That It Is a Complex Adaptive System

The complexity of the many linkages and feedbacks that make up a social-ecological system is such that we can never predict with certainty what the exact response will be to any intervention in the

---

BOX 3 Cogworld versus Bugworld

What's the difference between a complicated system and a complex adaptive system? Consider the situations of Cogworld and Bugworld.

Everything in Cogworld is made of interconnected cogs; big cogs are driven by smaller cogs that are in turn driven by tiny cogs. The size and behavior of the cogs doesn't change over time, and if you were to change the speed of the cogs of any size there is a proportionate change in the speed of other connected cogs. Because this system consists of many con-nected parts some would describe it as being complicated. Indeed it is, but because the components never change and the manner in which the sys-tem responds to the external environment is linear and predictable, it is not complex. It is just a more complicated version of a simple system, like a bicycle with multiple gears.

Bugworld is quite different. It's populated by lots of bugs. The bugs interact with each other and the overall performance of Bugworld depends on these interactions (as does Cogworld). But some subgroups of bugs are only loosely connected to other subgroups of bugs. Bugs can make and break connections with other bugs, and unlike the cogs in Cog-world, the bugs reproduce and each generation of bugs come with subtle variations in size or differences in behavior. Because there is lots of varia-tion, different bugs or subgroups of bugs respond in different ways as conditions change. As the world changes some of the subgroups of bugs perform better than other subgroups, and the whole system is modified over time. The system is self-organizing. No one is in control.

Unlike Cogworld, Bugworld is not a simple system but a complex adap-tive system in which it's impossible to predict emergent behavior by understanding separately its component subgroups.

The three requirements for a complex adaptive system (see Levin 1998) are:

---

system. Complex adaptive systems have emergent behavior; that is, the emergent behavior of the system cannot be predicted by understanding the individual mechanics of its component parts or any pair of interactions (see Levin 1997 and box 3, "Cogworld vs. Bugworld"). What's more, emerging results from studying complex systems demonstrate that changes in one component can sometimes result in complete reconfigurations of the system; the system changes to a different stable state (or regime).

- That it has components that are independent and interacting
- There is some selection process at work on those components (and on the results of local interactions)
- Variation and novelty are constantly being added to the system (through components changing over time or new ones coming in)

In Cogworld there is a direct effect of a change in one cog, but it doesn't lead to secondary feedbacks. The cogs that make up Cogworld interact but they are not independent, and the system can't adapt to a changing world. Cogworld might function very efficiently over one or even a range of settings but it can only respond to change in one way— that is working all together. If the external conditions change so that Cogworld no longer works very well—the relative speeds of the big and little cogs don't suit its new environment—there's nothing Cogworld can do.

In Bugworld the system adapts as the world changes. There are secondary feedbacks—secondary effects of an initial direct change. The bugs of Bugworld are independent of each other though they do interact (strongly—though not all bugs interact with all other bugs).

In our Bugworld, if we attempted to manage a few of the subgroups— that is, hold them in some constant state to optimize their performance— we need to be mindful that this will cause the surrounding subgroups to adapt around this intervention, possibly changing the performance of the whole system.

Ecosystems, economies, organisms, and even our brains are all complex adaptive systems. We often manage parts of them as if they were simple systems (as if they were component cogs from Cogworld) when in fact the greater system will change in response to our management, often producing a raft of secondary feedback effects that sometimes bring with them unwelcome surprises. The real world is more Bugworld than Cogworld—and yet it seems most of our management is based on the Cogworld metaphor.

As discussed, managing a system component-by-component can work well over short timeframes but it inevitably leads to long-term problems. Usually, the predictable short-term behavior of a particular component leads to the formulation of management guidelines that fail in the long term.

Commercial fisheries are a good example. Fishery managers set catch quotas for "maximum sustainable yields" based on the information they have gathered on the impact of fishing on the size of the target fish population. The aim is to get the fish population to the size where its reproduction (sustainable yield) is at a maximum. They assume that in response to harvesting, the size of future populations will behave as they have in the past, and that changes will be linear and incremental. Under this theory, changes in the harvesting pressure should result in corresponding shifts in the population stock.

Despite a high degree of confidence in the validity of this approach, commercial fisheries are failing all over the world as wild fish populations decline catastrophically and then do not recover even when fishing is stopped. The problem stems from the fact that the predicted harvest–response relationship has invariably been determined over a very limited range of population sizes, and the shape of the relationship across the full population size is unfortunately not a nice smooth one (Hilborn and Walters 1992).

Besides being inherently unpredictable, complex adaptive systems can also have more than one "stable state."[*] A change in a system can move it over a threshold into a different "stability regime," sometimes called an "alternate stable state"(Scheffer et al. 2001). A social-ecological system based on a wild fishery, for example, can cross a threshold and experience a catastrophic collapse in fish numbers. The fishing then stops but the fish population does not recover. The system has moved to a different stable state, a state in which the commercial levels of the fish population are absent. (The notion of multi-stable states and thresholds are discussed in greater detail in chapter 3.)

---

[*]The use of the term "stable state" here actually refers to many possible states a "stable" system can exist in. This set, or configuration, of states is constrained by the variables that govern the system. Many ecologists (cf. Scheffer et al. 2001) now use "regime" instead of stable state as it better describes the set of states within which a system tends to stay. These distinctions take on a greater significance as the building blocks of resilience thinking are explored.

Concept 3: Resilience Is the Key to Sustainability

The two concepts we have discussed so far lay down the context of systems thinking. First, systems are strongly connected (though everything is not connected to everything else) and we are part of the system; and, second, complex systems behave in nonlinear ways and are largely unpredictable. They can exist in different regimes (stability domains).

The third concept describes the property of a social-ecological system that an increasing number of scientists believe is the cornerstone of sustainability—resilience. Resilience is the capacity of a system to absorb disturbance without shifting to another regime (Holling 1973; Walker et al. 2004). A resilient social-ecological system has a greater capacity to avoid unwelcome surprises (regimes shifts) in the face of external disturbances, and so has a greater capacity to continue to provide us with the goods and services that support our quality of life.

Resilience in and of itself is of course not necessarily desirable. A social-ecological system in an undesirable state (such as a depleted fishery, a salinized landscape, or a murky lake system) may also exhibit high resilience and resist all efforts of managers to move the system out of that state. Think also of social states like Franco's Fascist regime in Spain. His dictatorship was remarkably resilient, it lasted from 1936 until 1975, despite sweeping worldwide changes.

Resilience means different things to different people. One interpretation is the capacity of something or someone to bounce back to a "normal" condition following some shock or disturbance. In this book we use the term with a different emphasis. Though the time it takes to get back can be important, resilience is not about the speed of the bounce back so much as the ability to get back. (This is an important distinction and is discussed in greater detail at the end of chapter 3.)

The case study we present on the coral reefs of the Caribbean is a good example of this distinction. Coral reefs experience disturbances all the time, usually in the form of storms (often hurricanes). Healthy coral reefs reassemble after the disturbance of a storm. They recover. Sometimes they recover quickly, sometimes more slowly. The speed of recovery is not so important. What is important is that the reef is *able* to recover, and that after the disturbance the direction it takes is back toward something like what it was.

But the coral reefs of the Caribbean have not been recovering over the last thirty years. They have lost this capacity due to a range of human interventions. They have crossed a threshold into a new regime.

In the Everglades, resilience is reflected by the marshland's capacity to absorb the repeated disturbances of flood, drought, and fire. Nutrients introduced into the system by agriculture, however, reduced the resilience of the sawgrass regime. Now in many parts of the Everglades, phosphorus-loving cattail is displacing sawgrass (and the habitat it provided for many species of water birds). The system is no longer absorbing disturbances as it was once able to, and has shifted into a new (cattail) regime.

The Goulburn-Broken Catchment, the subject of our next case study, is a rich agricultural region in southeast Australia. It is also a social-ecological system with multiple regimes, and it is changing. In the face of this change local industries are trying to become more efficient in order to stay viable (which is very much an optimizing approach). More of the same, however, may not be enough to save the region.

Key Points on Resilience Thinking

- When considering systems of humans and nature (social-ecological systems) it is important to consider the system as a whole. The human domain and the biophysical domain are interdependent. To consider one in isolation of the other is to come up with a partial solution that can lead to bigger problems down the line.
- Social-ecological systems are complex adaptive systems; understanding how their component parts function doesn't mean you can predict their overall behavior.
- Resilience thinking provides a framework for viewing a social-ecological system as one system operating over many linked scales of time and space. Its focus is on how the system changes and copes with disturbance. Resilience, a system's capacity to absorb disturbances without a regime shift, is the key to sustainability.

Between a (Salt) Rock and a Hard Place:
The Goulburn-Broken
Catchment, Australia

The dairy farmers of the Goulburn-Broken Catchment are praying for rain. Along with many other parts of Australia, they are experiencing one of the worst droughts on record and if it goes on for much longer many of them will go broke, drowning in unserviceable debt incurred as they buy feed for their herds.

They desperately need rain, but not too much rain. A string of really wet seasons might also spell doom for them as it could bring the water table, which lies perilously close to the surface, up to the root zone of their pastures. This would have catastrophic consequences because, besides water logging the soil, the groundwater carries with it a vast load of ancient stored salt that will effectively destroy the productivity of the land.

And if drought and flood weren't worrying prospects enough, they also can't afford for market conditions to change too much or too quickly. These are some of the most efficient farmers in the world employing state-of-the-art technology and producing high-quality, low-priced milk in large quantities. However, after decades of squeezing every ounce of efficiency out of their production systems, there's not a lot more they can do.

Indeed when it comes to water use, they can't afford to get any more efficient. Most of the dairying is done on irrigated pasture and they need the irrigated water they add to their pastures to flush the salt down past the root zone and out of the top two meters of soil to keep it productive. The pastures take up water leaving the salt behind, and if they reduce irrigation any more the salt will accumulate. But, if the groundwater rises into the top two meters of the soil, it, and the salt, will be drawn to the surface through capillary action.

So, they can't increase irrigation either, because this would cause the groundwater level to rise. The only thing that stops water tables rising is pumping. The farmers are hooked on their current levels of water use and levels of productivity with pumping of rising groundwater a permanent part of their lives. If anything changes too much there's a high price to pay.

All in all, we're looking at a highly productive region that has frighteningly little resilience. The farmers, and the local communities that depend on the prosperity of the farmers, are between a rock and a hard place. They have little room to move. They can continue to function as long as there are no major shocks to the system—and the current drought they're experiencing might be the straw that breaks the camel's back. And if that doesn't do it, the next run of wet seasons may.

How did this situation arise? It didn't happen because of a lack of resources, information, or commitment. The Goulburn-Broken Catchment (GBC) is one of the nation's most productive agricultural heartlands. It's also among the most intensively studied catchments. More is known about its biophysical function than just about anywhere. Knowing more hasn't helped because the underlying expectation of the people in the

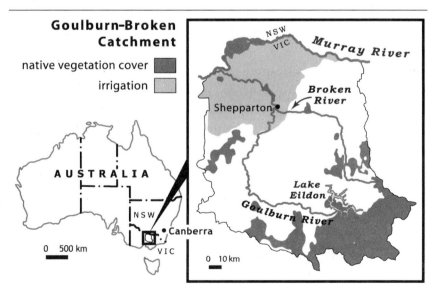

FIGURE 2 Map of the Goulburn-Broken Catchment

region is that they want to continue doing things the way they've always done things. Consequently they have thus far opted to fix up short-term problems rather than address the larger systemswide issues. And this situation hasn't arisen overnight. It's been approaching for decades.

So, what's the story? The root cause of the problem has been the radical alteration of the natural ecosystem, specifically its hydrology, through the clearing of most of the native forests and woodlands in order to farm. Resilience thinking allows us to interpret what's been happening.

The Goulburn-Broken Catchment in a Nutshell

When Major Thomas Mitchell, one of the first European explorers to visit the region, first looked out over the riverine plains in 1838 he spotted its rich potential immediately:

> I ascended a rocky pyramidic hill . . . the view was exceedingly beautiful over the surrounding plains, shining fresh and green in the light of a fine morning.
>
> The scene was different from anything I had ever before witnessed, either in New South Wales or elsewhere. A land so inviting, and still without inhabitants! As I stood, the first European intruder on the sublime solitude of these verdant plains, as yet untouched by flocks or herds; I felt conscious of being the harbinger of mighty changes; and that our steps would soon be followed by the men and the animals for which it seemed to have been prepared. . . . This seemed to me a country where canals could answer the better distribution of water over the fertile plains.

And his vision was realized.

The Goulburn-Broken Catchment (GBC) is a subcatchment of the Murray-Darling Basin, Australia's largest and most important river system. The Goulburn and Broken rivers rise in the southern high regions, join in the lower catchment and then flow into the Murray River, along the catchment's northern boundary. The catchment is about 120 by 200 kilometers with rainfall ranging from 1200 millimeters in the higher southeast regions to 400 millimeters in the northwest.

Although it is only 2 percent of the land area of the Murray River Basin, the GBC contributes 11 percent of the Murray's water flow. Salinity is a major threat to both the long-term viability of the region and to water users downstream. The catchment is a major contributor of salt loads to the Murray River, with total loads expected to increase substantially in the future as the hydrological system moves toward a new equilibrium.

The most productive area in the GBC is 300,000 hectares of irrigated country centered on the riverine plains in the lower catchment. Most of the water comes from a dam (Lake Eildon) in the upper catchment. This irrigated area produces around $1 billion Australian dollars worth of produce annually, with value adding and processing of produce generating a further AUS$3 billion. The region accounts for 25 percent of Victoria's export earnings, and is home to 190,000 people.

Native vegetation cover in the region has been reduced by about 70 percent since European settlement. The remaining 30 percent is largely in the mountainous terrain in the upper catchment and in small strips along waterways and roadsides. In the irrigation region there is less than 2 percent of native vegetation cover.

The Devil in the Landscape

The devil that is haunting the farmers of the GBC is the same one that is challenging irrigation farmers all around the world—the threat of rising salt.

Salt occurs naturally at high levels in the subsoils of most Australian agricultural land. It originally came from the ocean, carried inland on prevailing winds and deposited in small amounts with rainfall and dust. The vegetation absorbed and transpired the water, but left the salt behind in the soil. Rain would then flush this salt down into the subsoil. Over tens of thousands of years, however, these small amounts accumulated into vast stores of salt in the subsoils—up to fifteen thousand tons per hectare in some places.

Prior to European settlement, groundwater tables in Australia were in long-term equilibrium, deep below the surface. Native Australian vegetation commonly has deep root systems so plants can survive drought periods. The vegetation has developed such that the diverse mix of species fully exploits the annual input of rainwater in the soil. Evergreen native plants extract water from the whole soil profile, all year, preventing much water from entering the groundwater. Only in very wet years is the mass of roots unable to take up all the water, and in those years water flows down below the roots to the groundwater. But even if flooding rains were experienced over several seasons the groundwater was deep enough that it could rise without coming close to the surface.

With European settlement this equilibrium shifted. The settlers cleared most of the native vegetation and replaced it with annual crops and pasture species. The water balance changes when shallow-rooted pastures or crops replace deep-rooted perennial vegetation. These shallow-rooted plants, many of which grow for only a few months, do not use all the rain that falls so more water penetrates through to the groundwater. As a result, groundwater tables have risen, bringing with them the dissolved salt.

"Taming" the Goulburn Broken

The first occupants of the region were Aborigines. They have been present in the region for at least eight to ten thousand years (and probably much longer). The land was covered by open grassy woodland maintained by periodic fires. The population densities of Aboriginal people in the lower catchment, concentrated along the rivers, wetlands, and streams, were among the highest in Australia, testament to the area's high natural productivity. Back then the water table was somewhere between twenty to fifty meters below the surface.

The Europeans who followed Major Mitchell saw the rich riverine soils of the lower catchment as an excellent place to farm, and land began to be cleared soon after settlement in the late 1830s. The proximity of the mighty Murray River opened up the rich possibility of irrigation and intensive agriculture but the inherent variability of Australia's climate, a land of "drought and flooding rain," meant that any development was being severely tested by one disaster after another. A more reliable source of water was needed.

Broad-scale irrigated agriculture commenced in the lower GBC in the 1880s. Back then, the water table under this irrigated area was typically some twenty-five meters down. Small scale-irrigation systems along the major rivers were initially developed and managed by private water user associations, but these operations quickly failed due to mismanagement and the vagaries of drought. They depended on water being pumped directly from the river but when the river dried up, as it frequently did, so too did the irrigation.

Calls were made for bigger and better infrastructure to drought-proof

IMAGE 2

An aerial view of river flats in the Goulburn-Broken Catchment. This is the most productive land in the region, and most of the landscape has lost its cover of native vegetation. (*Courtesy of Paul Ryan.*)

agriculture in the area, and responsibility for the development and administration of a regional scale irrigation system was transferred to a centralized state bureaucracy.

The first major piece of infrastructure to be constructed to regulate the area's water was the Goulburn Weir in 1887. But this was followed by a series of the worst droughts on record and there was even greater demand for the construction of even bigger dams. The Eildon Dam was established in the upper catchment in 1916. Extensions to this dam were made in 1935 and 1950, each bout of construction following a debilitating drought. The 1950 extension increased the storage capacity of the dam by a massive 700 percent and was believed to guarantee the region's future prosperity, which was now grounded in intensive irrigation-based dairy and horticulture.

And the gods appeared to be smiling. A combination of positive terms of trade, strengthening individual property rights, subsidized pricing of irrigation water, and a period of exceptionally high rainfall between 1950 and 1960 meant boom times for the region. Such was the amount of rain received in this period that when the Eildon Dam extension was completed in 1956 it took only one year to fill instead of the expected five.

The good times also encouraged farmers to invest heavily in the expansion of irrigation. Things were going well and they progressively locked themselves into this ongoing development through their investment.

Releasing the Devil

While the farmers and government agencies were effectively managing the day-to-day variables of supplying irrigation water and growing produce, nobody was paying close attention to the one variable that effectively constrained the whole system—the rising level of the groundwater table. Although minor areas of high water tables were evident as early as 1940, these outbreaks were highly localized, occurring across less than 1 percent of the region.

The lower catchment had now lost all but 3 percent of its cover of native vegetation and was experiencing significant increases in water infiltration through irrigation. Indeed, the amount of water brought in for irrigation was the same as that falling as rain; rainfall, in other words, had effectively been doubled.

The system was on a new trajectory of rising water tables. The original depth to the water table had served as a buffer that had protected the region from the wide fluctuations of high rainfall periods, but as the water rose this buffer was rapidly being used up. The wet phase of 1950 to 1960 effectively removed that buffer completely, as water tables rose to within five to six meters of the surface. The lower catchment had effectively lost its resilience to anymore wet period shocks. The closer you are to a threshold, the less of a shock it takes to put you over it (see chapter 3).

Though the people of the catchment didn't know it, the good times of the 1950s had set them up to be completely exposed to the next extended wet phase which, as it turned out, occurred between 1973 and 1977. In previous years the extra rain that was received over that period would have been easily absorbed by the landscape. This time, however, it produced a crisis.

The shallow water table rapidly rose into the critical upper two-meter zone across more than a third of the region. It reduced production of dairy pastures and destroyed many high-value horticultural crops with 30 to 50 percent of the stone fruit crops being lost. A close examination

IMAGE 3

The wealth of the Goulburn-Broken Catchment. (*Courtesy of Paul Ryan.*)

of the situation revealed that more than half of the irrigation region (274,000 ha) was at risk of high water tables.

Dairy and horticultural production and processing, underpinning half the regional economy, were under threat and this had a profound impact on local communities, and beyond.

Crisis and Response

Crisis has a wonderful way of galvanizing action. The response to this rising water table crisis was on many levels. The immediate response was to install groundwater pumps and draw the water tables down to protect the fruit trees.

However, since the pumped water was discharged into the Murray, this action merely served to pass the problem up to the scale of the Murray-Darling Basin. Unfortunately, but not unexpectedly, the wider basin was also experiencing deteriorating water quality with the Murray River becoming increasingly salty, and there was a need to coordinate actions across the many subcatchments. The River Murray Commission was extended to become the Murray-Darling Basin Commission in 1986—a state/federal government partnership agency with powers over what was happening on the land as well as on the river.

IMAGE 4

Irrigation channels deliver the lifeblood of the region. (*Courtesy of Paul Ryan.*)

The Goulburn-Broken Catchment was now legally part of a larger-scale water management system, and was required to meet limits on how much salt could be exported to the Murray.

Community leaders in the catchment recognized that the water table crisis required a coordinated response. Action by individual farmers could not solve what was a common-resource problem so new community groups and networks came into being. "Landcare" groups were established that worked with a broad ethic of land stewardship, unlike previous single-focus community groups.

This was a revolution in how natural resource crises were dealt with in Australia. Previously, crisis management revolved around government control and intervention with little real input from regional communities. But this was changing and state and federal governments were beginning to relinquish some of their power. Coinciding with this was a national-scale movement of privatization. The net result was an empowerment of regional communities to drive their own decision-making.

A coalition of community and industry leaders lobbied at state and federal levels for assistance to address what they termed the "underground flood." Community leaders proposed a radical model of integrated catchment management based on community decision making,

IMAGE 5

A landscape damaged by rising salt. (*Courtesy of Paul Ryan.*)

which was in due course accepted. The state government had devolved responsibility for catchment management to regional communities in the form of Catchment Management Authorities (CMAs). The Goulburn-Broken Catchment Management Authority was one of the first CMAs to be created and served as a model for many subsequent CMAs being established around Australia.

The formation of the Murray-Darling Basin Commission also had major implications for the Goulburn-Broken as it established links that connected institutions operating at different levels of geographical scale.

A New Era

None of this institutional reform would have occurred without the need for change being catalyzed by the crisis of the underground flood in the 1970s. And the reforms that have been achieved in the Goulburn-Broken have been used as a benchmark for similar change in other regions and catchments around Australia. Some believe this was the beginning of a new era of land stewardship around the nation with a priority now being placed on local engagement, capacity building, and activity.

And yet, despite thirty years of effort, the Goulburn-Broken is still on the same trajectory it was on in the 1970s when the first crisis hit, locked into a losing battle with rising groundwater and rising salinity. Just as a social threshold has been crossed in the manner in which a region's natural resources are managed, a biophysical threshold has also been crossed and the catchment is now moving toward a new equilibrium. It exists in a new ecological regime—one it has actually been in since the early 1900s (Anderies 2005).

This new regime has different feedbacks and a new equilibrium level for the water table. Recent modeling of the catchment has reached a disquieting conclusion that the new equilibrium for groundwater tables is actually at the surface for much of the irrigated area and, without pumping, the amount of native vegetation that would have to be reestablished to move the catchment back to an equilibrium level of below two meters is around 80 percent cover. Further, the threshold of native vegetation cover that separated the old regime from the new regime was probably crossed about a hundred years ago (Anderies et al. 2005).

If we include pumping as an integral part of the new system, then the cover of native vegetation required drops significantly. However, pumping can't replace vegetation entirely due to a range of constraints.

An interesting hypothetical: if the early settlers had been forewarned of the problems that would be faced some one hundred years later, and been in possession of the information we now have, would they have made different decisions on how they developed the region? Probably not. A delay of one hundred years between an action and its consequences makes it difficult to take those consequences seriously. Humans have both high discount rates and an enormous capacity in believing the future will generate solutions to problems that don't have to be faced in the foreseeable future.

Just Coping

If the current situation in the GBC were to be left unmanaged, vast tracts of low lying land, including some of the region's most productive areas, would quickly become waterlogged, salinized, and unproductive.

The crisis of the 1970s woke everyone up to the potential threat of rising water tables. However, even with a large and concerted response, the result has only been to lower the water table in many areas to just out of reach of the root zone. This has been achieved by physically pumping out of the ground around 100 million liters of water every year via some five hundred water pumps. Some of this water isn't yet too salty and can be used again as irrigation water (which means a proportion of it going back into the groundwater eventually turning it salty). Some of it is flushed down the Murray River but a "salt cap" limits the amount that can be disposed of this way.

Pumping out water is expensive. While it can be currently justified to save the highly productive irrigation areas, it's still not a sustainable solution in the longer term. Reusing some of the pumped out water for irrigation will become increasingly difficult as salt levels rise. Setting up more evaporation basins is also expensive and raises a raft of equity issues (who will pay for them, and who will lose land to allow them to be established?), and a loss of amenity (it is mostly a very ugly land use).

Revegetation with deeply rooted crops and pastures and the widespread reestablishment of native vegetation cover are an important part of the longer-term solution. However, revegetation is a slow process and there is a significant lag between the time revegetation occurs and the time when groundwater-controlling benefits are realized. The scale and the cost of this revegetation, most of which needs to occur on privately owned land where the owners are either unwilling or unable to assist, is so high that little progress has been made on this score since the crisis in the 1970s. In many areas it's not even an option as the salt will concentrate under the growing trees (as the trees transpire just the water) retarding their growth and eventually killing them.

At the moment the catchment is simply coping; and its ability to cope has a lot to do with the catchment experiencing a prolonged dry phase. It is caught in a regime in which it doesn't want to be but from which it can't easily escape. Its best efforts are just keeping the salt devil at bay. It is not increasing the size of its safety buffer, but the costs keep on mounting and salt concentrations are accumulating. It lacks resilience and is vulnerable to environmental and economic shocks.

Looking Ahead

Our best guide to what climate might be expected in the coming decades is what has been experienced in the past. The climate of the Goulburn Broken is characterized by periodic droughts with intermittent wet periods. The current prolonged dry spell has reduced the immediate threat from rising groundwater, but is stressing the economy so the locals don't have large financial resources in reserve. There isn't enough stored water for growers to get their usual water entitlement, and so smaller acreages are irrigated.

When it does again rain, the catchment will likely cope with a single wet year. However, with several wet seasons in a row it likely won't be able to pump its way out of trouble. The groundwater will rise into the danger zone and the resulting damage will be much worse than that experienced back in the 1970s because the groundwater level is now higher across a greater area. The catchment is full; the water has nowhere else to go.

And the region is not just vulnerable to biophysical shocks. It has pinned its hopes on a narrow set of commodities. The changing economic environment means the catchment's enterprises have limited capacity to absorb shocks. Horticulture and dairy are becoming increasingly vulnerable within a changing global market.

The triumph of the region's response to the crisis of the 1970s was its capacity to engage its own community and create the local networks and institutions that have enabled it to sustain its productive base and regional vitality. It developed high adaptability. The flip side of that same response was a failure to acknowledge fully the underlying cause of the problem and to begin to explore alternative futures for the region. It failed to enhance its transformability.

Rather than consider the alternatives, all efforts were put toward getting back to business as usual. Working smarter and harder was thought to be the solution. In fact it was only reinforcing the problem. This failure has set the region up for even bigger problems down the line.

Many people are looking at the Goulburn-Broken to see how it fares over the coming years. Other catchments are facing equally dire problems with rising groundwater and salt. Given the resources and effort that have gone into the Goulburn-Broken following the crisis of the 1970s, if they can't make it work there then some believe it may not work anywhere.

The immediate options appear to be some combination of switching from the highest water uses (pasture for dairy) to lower uses (horticulture), using the water available for irrigation to keep flushing the soil but significantly reducing total input to the groundwater; revegetation of significant areas of the catchment; more pumping; and the development of novel forms of high-value land use that do not require irrigation. Small changes won't be enough. The region needs to transform the manner in which it functions, something that is never easy. Unfortunately, the longer it takes to make the transformation, the higher the cost of that process.

Some community leaders suspect that, after the next big wet, the new steady state for the region will probably involve about one third of it being salinized and another third being periodically subjected to water at the surface, but low in salt. Some believe they may be able to work with this. Pasture can be replanted after periodic inundation as long as irrigation is used afterward to flush salt out of the top layers of soil. Trees (fruit crops and native vegetation) are not so hardy and would be killed by the rising groundwater, so the region would become progressively more dependent on dairy. But this would be a future with even less resilience.

The way forward is not clear or easy. A sustainable future involving transformation will depend on how much land is revegetated, the pattern of wet periods that may be experienced while this vegetation is establishing and the region's capacity to diversify its economic activity in completely novel (non-irrigation) ways. Yet, there is reason to hope. The community is aware of the problems and has developed a significant capacity to work together. That capacity might be the critical factor in this social-ecological system's resilience as it faces the future.

Resilience and the Goulburn-Broken Catchment

This Goulburn-Broken story demonstrates the critical importance of understanding the underlying variables that drive a social-ecological system, knowing where thresholds lie along these variables, and knowing how much disturbance it will take to push the system across these thresholds. To ignore these variables and their thresholds, to simply focus on getting better at business as usual, is to diminish the resilience of the system, increase vulnerability to future shocks (droughts, wet periods, and economic fluctuations) and reduce future options. Being more efficient is not by itself a pathway to sustainability. Because resilience was not being consciously factored into the management of the region, greater production efficiency has actually reduced the possibilities of the system being sustainable.

3

Crossing the Threshold:

Be Careful about the Path You Choose—

You May Not Be Able to Return

Thresholds are levels in controlling variables where feedbacks to the rest of the system change—crossing points that have the potential to alter the future of many of the systems that we depend upon. They are all around us, however we often aren't aware of them until after they've been crossed and we observe that the system is behaving in a significantly different manner. As we saw previously, the social-ecological system of the Goulburn-Broken Catchment has crossed a threshold. It now behaves in a different manner. The ramifications of this are still unknown but it has undoubtedly limited the future options of the people who live there.

The framework for resilience thinking is based on two ways of seeing and understanding social-ecological systems. One is based on a metaphor of adaptive cycles (explored in chapter 4) and the other focuses on the likelihood of a system crossing a threshold and moving into a different regime. This thresholds model is usefully described with the metaphor of a ball in a basin.

A System as a Ball in a Basin

The important variables you use to describe a system are known as the system's "state" variables. If the system consists of the number of fish and the numbers of fishers, you have a two-dimensional system. If the system consists of the amounts of grass, trees, livestock, and people employed in ranching, it is four-dimensional.

We can envisage the system as a number of basins in two-, or four-, or

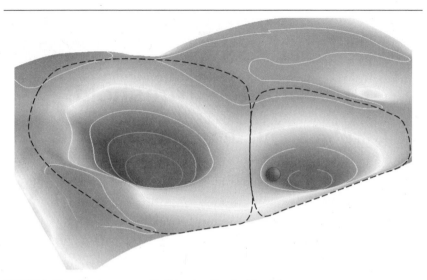

FIGURE 3 The System as a Ball-in-the-Basin Model

The ball is the state of the social-ecological system. The basin in which it is moving is the set of states which have the same kinds of functions and feedbacks, resulting in the ball moving towards the equilibrium. The dotted line is a threshold separating alternate basins. *(From Walker et al., 2004.)*

n-dimensional space as depicted in figure 3. The ball is the particular combination of the amounts of each of the n variables the system currently has—that is, the current state of the system. The state space of a system is therefore defined by the variables that you are particularly interested in, encompassing the full array of possible states the system can be in.

Within a basin (where the system has essentially the same structure and function, and the same kinds of feedbacks) the ball tends to roll to the bottom. In systems terms, it tends toward some equilibrium state. In reality, this equilibrium is constantly changing due to changing external conditions; however, the ball will always be moving toward it. The net effect is that one never finds a system in equilibrium (i.e., with the ball at the bottom of the basin). The shape of the basin is always changing as external conditions change (see figure 4 and also Scheffer et al. 2004) and so is the position of the ball. So the system is always tracking a moving target and being pushed off course as it does so. From a resilience perspective the question is how much change can occur in the basin and in the system's trajectory without the system leaving the basin.

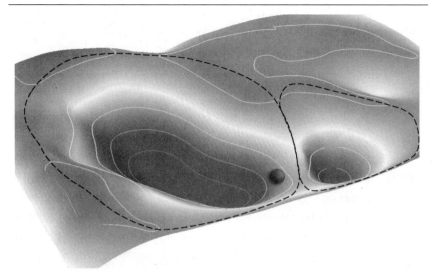

FIGURE 4 The Basin Changes Shape

This this is the same system as in figure 3. The state of the system (position of the ball) has not changed, but as conditions change, so too does the shape of the basin and the behavior of the system. *(From Walker et al., 2004.)*

Beyond some limit (the edge of the basin), there is a change in the feedbacks that drive the system's dynamics, and the system tends toward a different equilibrium. The system in this new basin has a different structure and function. The system is said to have crossed a threshold into a new basin of attraction—a new regime. These differences can have important consequences for society and so some basins of attraction are deemed desirable and others not.

And it's not just the state of the system (the position of the ball) in relation to the threshold that's important. If conditions cause the basin to get smaller, resilience declines, and the potential of the system to cross into a different basin of attraction becomes easier. It takes a progressively smaller disturbance to nudge the system over the threshold. Figures 3 and 4 shows this using the ball in the basin analogy.

Picture a Lake

A good example of a threshold being crossed is what can happen to a lake that receives plant nutrients, such as phosphorus, in runoff from the surrounding land (based on the work of Carpenter 2003 and Scheffer et al. 2001). Over time the level of phosphorus in the lake water increases, and it accumulates and gets stored in the lake sediments.

As most people know, phosphorus is a plant nutrient and encourages the growth of algae. This algal growth can turn clear water murky. After a runoff event, such as might be triggered by a big rainstorm, phosphorus levels in the water go up, algal growth increases and the clear lake goes murky (see figure 5). The amount of phosphorus in the water depends on two sources: the phosphorus in the water that is flowing in, and the amount of phosphorus in the lake sediments. If the phosphorus in the sediments is low then the sediments will absorb phosphorus in the water. When the water near the sediment has plenty of oxygen, the phosphorus gets bound up in the sediment in a form that has low solubility; little gets released back into the water. The algae therefore lose their source of nutrients, and the lake returns to a condition of clear water.

If the phosphorus in the lake sediments is low, it draws the phosphorus levels in the water down, and so the lake has a degree of resilience to shocks caused by nutrient inputs of phosphorus. It may initially turn murky from a shock of nutrient input but it will soon return to clear water. Its buffer is the amount of phosphorus that the lake sediments can store.

In terms of the ball in a basin metaphor, the ball is the state of the lake (basically, the amount of phosphorus in the water). The equilibrium (the bottom of the basin of attraction) is the clear water state (low

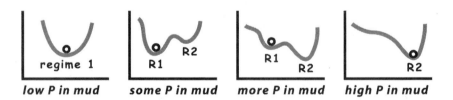

| *low P in mud* | *some P in mud* | *more P in mud* | *high P in mud* |

FIGURE 5 A Two-Dimensional Representation of a Ball-in-a-Basin Model of a Lake Ecosystem Changing over Time with Continued Phosphorus Inputs

Regime 1 (low phosphorus levels in the mud) is fully resilient to shocks of phosphorus inputs (as indicated by the depth of the single basin in which it sits). As phosphorus accumulates, the lake loses this resilience; the basin flattens and a new basin appears. Now the lake is vulnerable to a disturbance in the form of a rain event that suddenly raises phosphorus levels and eventually the system is easily pushed into a new basin of attraction (Regime 2), which is always murky. *(Adapted from Folke et al., 2004.)*

phosphorus and low algae). The size of the basin of attraction is how much phosphorus can be added to the water before it loses the ability to return to a clear water state. An increase in nutrients changes the position of the ball within the basin, but up to a point doesn't fundamentally alter the way the lake system behaves. The lake ecosystem can cope with a certain amount of phosphorus and still maintain its basic function. Its buffer is the distance to the threshold.

If over time phosphorus continues to accumulate in the lake sediments the potential for a new regime begins to develop. A threshold emerges. It happens like this: algal growth accumulates during a period of high phosphorus in the water. As dead algae rain down into the bottom layers they decompose and in so doing deplete the oxygen in the bottom waters; this is where a feedback change occurs. In low oxygen conditions the phosphorus in the sediments becomes very soluble and this leads to a sudden increase in phosphorus levels in the water—the more phosphorus there was in the mud, the more that gets released.

The system has now crossed a threshold into a new regime in which the feedback from phosphorus in the mud to released phosphorus in the water is many times higher—high enough to ensure that, even if no more phosphorus flows in from the surrounding land, the water phosphorus will remain sufficiently high to sustain algal growth, which in turn will keep the bottom water low in oxygen. The lake cannot recover to its former clear state. In this new regime, though the exact state may vary somewhat, the algal growth proceeds unchecked, the water becomes murky and smells, fish die, and the system no longer provides the ecosystem services it once did to the surrounding population.

Returning to the clear condition involves a hysteresis (or lag) effect— the sediment phosphorus must go much lower than the level where it became very soluble before water phosphorus again drops. Getting the phosphorus down to this level can take a long time (through gradual removal via leakage or harvesting) or be very expensive.

This murky (or eutrophic) lake regime is a new basin of attraction. In systems jargon it has a new attractor—a new stable state. The state of the system (position of the ball) is now controlled by a new set of feedbacks. Merely reducing the phosphorus input to the water (from the surrounding land) isn't going to restore the old regime.

There are a couple of aspects to reflect on here. The first is that in this lake example the system involves the surrounding human population and land use as well as the lake itself. It is a social-ecological system with a social as well as an ecological dimension. Once the threshold has been crossed there are consequences for both the social and the ecological components of the system.

Second, the amount of phosphorus that can be absorbed by the lake water before the system crosses into this new regime depends on the concentration of phosphorus in the sediment. As the amount of sediment phosphorus increases, so the system's resilience to phosphorus input decreases. The amount of extra water phosphorus (in the form of a pulse in run off from the surrounding land) required to push the system across the threshold into a different regime becomes smaller and smaller. Eventually, just a small phosphorus inflow (a small external disturbance) is all it takes.

Thresholds in the Real World

The process of a lake system receiving an excess of nutrients from surrounding land uses, crossing a threshold, and flipping into a different and persistent state is one that has been witnessed in many kinds of systems in many parts of the world (Walker and Meyers 2004).

Shagawa Lake in Minnesota, serves as one good example (Carpenter 2003). It suffered eutrophication from sewage inputs. The declining water quality of the lake prompted partial sewage treatment in 1911 but water quality continued to deteriorate. Treatment improvements in 1952 also failed to improve the quality of the lake water. Finally, in 1973, a new treatment plant was built which decreased phosphorus inputs by more than 80 percent and yet summer concentrations of total phosphorus remained unchanged. By 1991, the lake had still not recovered. Sometimes, when you cross a threshold there is no easy way back.

Eutrophication of a lake system is one example of a social-ecological system crossing a threshold. There are many others. The Goulburn-Broken Catchment is a case of a threshold involving the amount of native vegetation cover and the depth of the water table. The Everglades experienced increased nutrient run off from agricultural land that exceeded a threshold that has resulted in the spread of phosphorus-loving cattails across the northern Everglades, displacing the previously common sawgrass. In the Caribbean, overfishing and eutrophication changed the feedbacks for competition between corals and algae such that the coral cover no longer has the capacity to regenerate following disturbance events such as hurricanes. Brown algae now dominate in many areas and corals are in decline or completely lost in most areas of the Caribbean.

Slow Variables

The trajectory of our hypothetical lake system was determined by one key, slow-moving variable—phosphorus levels in the lake sediments. Though multi-scale social-ecological systems involve more than one variable, their trajectories are nevertheless governed by only a handful of interacting variables. If there were many more variables involved the system would be in a mode of constant regime shifts (state flips), each involving losses of resources (nutrients, organisms, species, etc.) and could not persist.

Managing for resilience is all about understanding a social-ecological system with particular attention to the drivers that cause it to cross thresholds between alternate regimes, knowing where the thresholds might lie, and enhancing aspects of the system that enable it to maintain its resilience. If you have some idea about the slow controlling variables in a system, and where thresholds lie along those variables, you have a head start.

If you think of a system as a ball moving around in a basin of attraction, then managing for resilience is about understanding how the ball is moving and what forces shape the basin. The threshold is the lip of the basin leading into an alternate basin where the rules change. The capacity of the actors in the system to manage resilience is described as adaptability. This might be by moving thresholds, moving the current state of the system away from a threshold, or making a threshold more difficult to reach.

In terms of the lake and phosphorus inputs example, this might translate into establishing belts of riparian vegetation around the lake that would filter inflowing water and reduce the phosphorus input to the lake. Or it might be possible to manipulate the food chain of the lake to boost the number of zooplankton (small animals suspended in the water) that consume algae. This might be done by reducing the number of fish in one layer of the "trophic cascade" or perhaps adding a top predator. Increasing the number of zooplankton would increase the threshold of the phosphorus in the water as there would be lower levels of algae for the same level of water phosphorus.

If the system is stuck in an undesirable basin of attraction then it

BOX 4 Thresholds Are Everywhere

Thresholds are levels in controlling slow variables after which feedbacks to the rest of the system change—crossing points that have the potential to alter the future of many of the systems upon which we depend. Ignoring them is a bit like playing Russian roulette—each new empty chamber passed is one step closer to the bullet—and yet few natural resource management schemes even acknowledge their existence.

The Resilience Alliance is building up a database of known and proposed thresholds. You can view it on their website (see http://resalliance.org). Below are just four examples. The Florida Bay is directly linked to the Everglades (case study 1) in that the bay receives most of its water after it has passed through the Everglades.

Easter Island

Easter Island was settled around 800 A.D. It was covered by a tropical forest, with 6 species of land birds and 37 species of breeding seabirds. The trees were felled for firewood, making gardens, building canoes, and for rolling and levering the giant statues carved on the island. By 1600, all of the trees, land birds, and all but one of the seabirds were gone.

At its peak, the island's human population grew to an estimated 10,000 people. The level of forest felling exceeded the rate of regeneration. At some point in forest decline the system moved into a new regime in which soil erosion was high and trees couldn't regenerate. The statues could no longer be constructed and (more significantly) canoes, used for catching large fish and marine mammals for food, could no longer be built. The society resorted to cannibalism and the population collapsed to an estimated 2,000 people.

Looking at Easter Island today it's hard to imagine the original lush forest cover.

Alternate Regimes

Tropical forest, land and sea birds
Eroded landscape, grassland (no trees), extinction of birds

Slow Variable

Nitrogen content of the soil. The threshold occurred in feedback to plant growth—when it was too low for tree regeneration the system shifted to a new regime. It was crossed because of forest clearing and burning.

Florida Bay

Florida Bay is a shallow lagoon between the Florida mainland and the Florida Keys. Its western margin opens to the Gulf of Mexico. It receives freshwater runoff from the land and canal systems, and water salinity fluctuates season-ally from brackish to hyper-saline. More than 4000 ha of sea grass beds were completely lost between 1987 and 1991 and a further 23,000 ha were affected to a lesser degree. There was some recovery after 1992.

Land use changes surrounding Florida Bay have altered the water levels, increased salinity and lowered water circulation in the bay. Most likely, a combination of events contributed to the change of regime. The die-off occurred in parts of the bay that had less water circulation and consequently lower dissolved oxygen levels. The change to the eutrophic state was rapid and patchy. Recovery, which began in 1992, was much slower with algal blooms and turbidity plumes still observed after this date.

Clear water, healthy seagrass, and abundant fish species
Murky water, plankton blooms, few fish

Slow Variable

Cover of seagrass. It's believed that increased nutrient input, or changes
in water depth, temperature, hurricane frequency, and/or salinity led
to decreases in sea grass cover.

Sea Otters and Marine Ecosystems of the Pacific Rim

Once distributed across the Pacific Rim, sea otters were hunted to near
extinction in the eighteenth and nineteenth centuries. Less than 1000
remained by the early 1900s in the Aleutian Archipelago.

Sea otters feed on sea urchins and shellfish, keeping their populations
in check. When some populations of sea otters were harvested to extinction in the Aleutian Archipelago, sea urchin populations grew unchecked.
The sea urchins grazed heavily on kelp leaving little food or shelter for
fish, which declined in number. In turn, the harbor seals, which feed
predominantly on fish, also declined in number. The threshold density of
sea otters required to shift the system back to Regime 1 is unknown. Also
unknown is the threshold density of sea urchins that can be sustained
before the numbers of kelp, fish, and harbor seals declines.

Alternate Regimes

High density of sea otters, low density of sea urchins, high density of
  kelp, fish, and harbor seals
No sea otters, high density of sea urchins, low density of kelp, fish,
  and harbor seals

Slow Variable

Densities of sea urchins (slower) and sea otters (slowest). The threshold
relates to the level of hunting of sea otters.

Australia's Northern Savannas

Savannas (grasslands with low densities of trees) are important grazing
lands (rangelands). At low stocking levels, with periodic fires, the system
will remain grass dominated. As stocking rate increases, and especially
during periods of low rainfall, there is added pressure on the grasses,
which eventually decline to levels that cannot carry a fire. In the
absence of fire, woody vegetation becomes dominant, preventing grass
growth, and grazing productivity declines. Even without grazing, there
is not enough grass for a fire (sometimes referred to as a shrub desert).

As the older woody vegetation dies, and if there is a long period of low
intensity grazing, the gaps may again be colonized by grass. Fire plays a
major role in returning the system to grassland.

Alternate Regimes

Grass-dominated savanna
Shrub-dominated savanna

Slow Variable

Grazing pressure, level of rainfall. The thresholds relate to stocking rates
of cattle and fire frequency.

might be that it's impossible or too expensive to manage the threshold or the system's trajectory. In such cases it might be more appropriate to consider transforming the very nature of the system—redefining the system by introducing new state variables. Transformability is the capacity to create a fundamentally new system when ecological, social, economic, and political conditions make the existing system untenable.

A good example of this is what happened in Zimbabwe a couple of decades ago. In southeastern Zimbabwe cattle ranching was the dominant land use for many decades. But declining terms of trade and increasing amounts of woody shrubs had made it less and less profitable. In the early 1980s there was a severe, two-year drought that led to 90 percent of the cattle dying. However, ranchers noted that the remaining wildlife (that they had been busy eliminating in previous years) fared much better. Rather than persist with an enterprise that was firmly lodged in an undesirable basin of attraction, many landholders reinvented their enterprises (Cumming 1999). They joined their properties, removed internal fences, and transformed their farms into game safari parks. Their efforts met with enormous success (though subsequent political events at the national scale have once again created undesirable outcomes).

Thresholds Define Resilience

In chapter 2 we defined resilience as the capacity of a system to absorb disturbance, undergo change, and still retain essentially the same function, structure, and feedbacks—the same identity. At the end of this chapter it's worth reviewing this definition in relation to thresholds.

In everyday usage, most people think of resilience as the ability of someone or something to bounce back. When someone says, "He's a resilient child" or "this is a resilient community," they usually mean that he or the community have the capacity to get back on with life quickly after a shock or some disturbance.

The term "engineering resilience" is sometimes used to describe how quickly a system, often a mechanical system, can return to some point of equilibrium when disturbed. In a sense, therefore, this too is about bouncing back though this is more a measure of a system's stability.

Resilience as discussed in this book refers not to the speed with which a system will bounce back after a disturbance so much as the system's capacity to absorb disturbance and still behave in the same way. It's about what happens near the edge of the basin, not what happens near the equilibrium point at the bottom. Another way of saying that is: How much disturbance and change can a system take before it loses the ability to stay in the same basin? This is described as ecological resilience (Holling 1973).

Engineering resilience doesn't consider thresholds. Ecological resilience is all about identifying and understanding them. When you hear managers and planners using the term resilience (for example: "we are building a resilient industry" or "we are planning a resilient city") it is unclear which meaning they have in mind. Often they are talking about engineering resilience in which the aim is to bounce back quickly to business as usual following a small disturbance.

The distinction between "bouncing back" and "retaining the ability to get back" is crucial. Although the speed with which a farm, a catchment, or a region recovers after some disturbance can also be important, resilience thinking is much more about the ability of the system to recover at all.

The next case study on coral reefs is a good example of this. A healthy coral reef has the capacity to reassemble after a disturbance. However, the coral reefs in the Caribbean have lost this capacity, and that loss comes with a significant cost.

Key Points on Resilience Thinking

- Though social-ecological systems are affected by many variables, they are usually driven by only a handful of key controlling (often slow-moving) variables.
- Along each of these key variables are thresholds; if the system moves beyond a threshold it behaves in a different way, often with undesirable and unforeseen surprises.
- Once a threshold has been crossed it is usually difficult (in some cases impossible) to cross back.
- A system's resilience can be measured by its distance from these thresholds. The closer you are to a threshold, the less it takes to be pushed over.
- Sustainability is all about knowing if and where thresholds exist and having the capacity to manage the system in relation to these thresholds.

CASE STUDY 3

Losing the Jewel in the Crown:

The Coral Reefs of the Caribbean

The coral reefs are the jewel of the Caribbean crown. Threading their way along thousands of kilometers of coastline, the coral ecosystems not only provide food for millions of people, they also protect coastlines from the worst ravages of storms and create much of the sand for the region's beautiful beaches. Possibly their most important role, however, is their pulling power for a thriving tourism industry, the region's most important economic sector. It's difficult to think how an ecosystem could be of more direct value to its people.

It's estimated that the Caribbean coral reefs provided goods and services with an annual net economic value in 2000 of between $3.1 to $4.6 billion from fisheries, dive tourism, and shore line protection (Burke and Maidens 2004). And yet, for all its importance to the region's prosperity and future, the coral reefs of the Caribbean are in severe decline and most of the available evidence suggests it is people that are killing them.

Thirty years ago, the area of reef covered by hard coral was around 50 percent. Today it stands around 10 percent meaning there's been an 80 percent decline in cover of hard corals (Gardner et al. 2003). A recent report by the World Resources Institute quantified the extent of the threat posed by different activities (Burke and Maidens 2004). It has estimated that one-third of the region's reefs are threatened by pressures associated with coastal development, including sewage discharge, urban run-off, construction, and tourist development. Analysis of more than three thousand watersheds across the region identified 20 percent of the coral reefs were at high threat from increased sediment and pollution from

IMAGE 6

Hurricane Frances, the third major hurricane of the 2004 storm season, spins over the Caribbean. *(Courtesy of NOAA.)*

agricultural lands and other land modification. Fifteen percent of Caribbean reefs are threatened by discharge from wastewater from cruise ships, tankers, and yachts, leaks or spills from oil infrastructure, and damage from ship groundings and anchors. Overfishing threatens most Caribbean coral (we'll explain how, shortly), especially inshore reefs close to human population centers.

In addition to these quantifiable threats is the growing impact of disease and warming sea surface temperatures. Diseases have caused profound changes in Caribbean coral reefs in the past thirty years, with very few areas unscathed, even reefs far removed from human influence. In addition, coral bleaching episodes, associated with stress from climate change, are on the rise.

The Caribbean in a Nutshell

The Caribbean is not a single country or body of water but a vast watery domain that encompasses the Caribbean Sea, the Gulf of Mexico, and part of the northwestern Atlantic Ocean.

It is an area of high cultural and political diversity shaped by a vivid history. Within this region lie twenty-four sovereign nations (fourteen of which are island nations) and eleven territories of distant countries. These countries range from the world's richest nation (the United States) to some of the poorest; from long-established democracies to totalitarian regimes; and from industrialized countries with intensive agricultural systems to countries with little industry and largely natural landscapes.

Around 8 million square kilometers of land drain into the Caribbean region. This land stretches from the Upper Mississippi Basin in southern Canada to the Orinoco Basin of Colombia and Venezuela. Some 40 million people live within ten kilometers of the coastline. Coral reefs contribute significantly to nutrition and employment, particularly in rural areas and among island communities, where there are few employment alternatives.

Coastal shelves and warm tropical waters create ideal conditions for the formation of an estimated twenty-six thousand square kilometers of coral reefs. Separated from other coral reefs by distance and land barriers, they have evolved in isolation for millions of years, and remarkably few of the many thousands of species in these waters are found anywhere else in the world.

Over a third of the reefs are located within two kilometers of populated land. Tourism has surpassed fishing as the most important economic activity for many coastal localities.

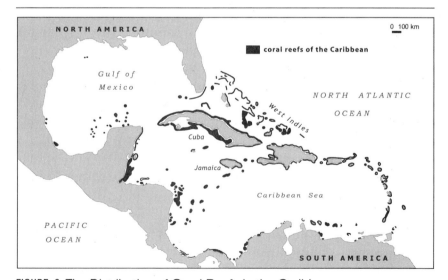

FIGURE 6 The Distribution of Coral Reefs in the Caribbean

IMAGE 7

Multiple impacts of runoff, pollution, global warming, over fishing, and tourism are all exacting a toll on the corals of the Caribbean. *(Courtesy of Lauretta Burke, World Resources Institute.)*

Coral Reefs and Resilience

First it was Charley, next it was Frances, then it was Ivan, and finally it was Jeanne; four major hurricanes in one season, making 2004 one of the costliest storm seasons on record. And, as they tore a path through the Caribbean, coral reefs were thrashed in the process. In centuries (even millennia) past, this wouldn't have mattered. Coral reefs have proved to be remarkably resilient structures.

Coral reef ecosystems exist in areas that constantly experience disturbances such as tropical hurricanes, and healthy reefs have the capacity to readily reassemble following such events. But in recent decades this hasn't been happening in the Caribbean. Instead, many reefs have gone from being dominated by hard corals to being dominated by fleshy seaweed (Bellwood et al. 2004).

The collapse of the coral reefs didn't happen overnight. It was long preceded by dwindling stocks of fish through overfishing, and increased nutrient and sediment run off from the land. Sediments smothered coral and nutrients and favored the growth of fleshy seaweed that was no longer being controlled by grazing fish. By the 1960s, the only thing preventing fleshy seaweed from taking over was an abundance of the *Diadema* sea urchin, which was also prospering from a lack of sea urchin–eating fish.

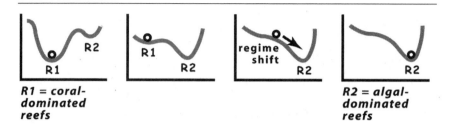

R1 = coral-
dominated
reefs

R2 = algal-
dominated
reefs

FIGURE 7 A Two-Dimensional Representation of a Ball-in-a-Basin Model of a Coral Ecosystem Changing over Time with Overfishing

Regime 1 (R1, dominated by corals) is resilient to shocks of storm damage (as indicated by the depth of the basin in which it sits). The algal regime (R2) is a possibility, but it has a very small basin of attraction. As functional groups of fish are removed through overfishing and nutrients build up through coastal eutrophication, the coral regime loses resilience (the basin flattens) and the algal basin (R2) grows. Now the coral regime is vulnerable to a shock and the state of the system is easily pushed into the algae basin of attraction which that is dominated by fleshy seaweed. (*Adapted from Folke et al., 2004.*)

In the 1970s, the density of *Diadema* on overfished reefs was extraordinarily high. But, while they kept the seaweed under control, their feeding action also served to erode the hard coral framework of the reef.

In the early 1980s a disease outbreak decimated the *Diadema* sea urchins throughout the Caribbean. The result was widespread blooms of seaweed that still persist. Today, the coral that remains is further affected by increasingly prevalent coral disease and coral bleaching.

On the other side of the world lies the Great Barrier Reef. Spanning an epic two thousand kilometers along the east coast of Australia, this reef is home to several thousand types of sea creatures. While it is suffering from its own set of pressures, at this point it has not witnessed the catastrophic declines that have been seen in the Caribbean. In part this is because it hasn't had the same intensity of fishing. However, the Great Barrier Reef also has a natural advantage—prior to any human influence, it had many more species of fish and coral than its Caribbean cousins.

The Caribbean went through a species bottleneck some 2 million years ago, thanks to a sudden sea level change and associated temperature shift. The result was a dramatic loss of species. The Caribbean reefs have only a fraction of the number of species found on the Great Barrier Reef, approximately 28 percent for fishes and a mere 14 percent for corals. The Great Barrier Reef has a much higher diversity.

IMAGES 8 & 9

Resilient coral reefs (as pictured on the left) have the capacity to rebuild following disturbances such as hurricanes. If they lose functional groups of fish they lose this capacity and the system moves into a regime in which the hard corals fail to regenerate (as pictured on the right). *(Courtesy of Terry Hughes.)*

Response Diversity

It's not just a matter of more species making for a more resilient system; it's the kinds of roles they perform. We need to distinguish here between "functional" and "response" diversity.

Functional diversity refers to the different functional groups of organisms that are represented in a system. By definition, different functional groups do different things in an ecosystem. One group might fix nitrogen, another might assist in the breakdown of waste, and another might provide the service of population control (by preying on a population of grazers and keeping their numbers in check). Within each functional group there is usually a range of species that provide the same basic service, though they go about their business in slightly different ways. For example, in one ecosystem there might be a number of different species that graze on fleshy algae.

When it comes to resilience, what's important is that the different organisms that form part of the same functional group each have different responses to disturbances (Elmqvist et al. 2003). In the case of the algae-eating species, resilience is enhanced if the different species, all of which are providing the same basic service of controlling the algae, each respond in different ways to changes in temperature, pollution, and disease. If there are a large number of different response types, the service provided by a functional group is likely to be sustained over a wider range of conditions, and the system has a greater capacity to absorb disturbances.

The range of different response types available within a functional group is referred to as response diversity, and it's this aspect of diversity that is critical to a system's resilience. It's akin to risk insurance and portfolio investment, something easily understood by anyone involved in business management. And, it's also what is limiting in many of the functional groups that maintain the Caribbean's coral reefs.

On coral reefs different functional groups of corals are often classified by the shape of their colonies. In the Caribbean, several functional groups of coral are missing or represented by only a handful of species (i.e., there is very limited response diversity). There are, for example, no three-dimensional bottlebrush species, and just one staghorn and one tall, tubular coral. Until recently, these two species were abundant and widespread, commonly comprising more than 30 to 50 percent of the total coral cover. Today, many areas have lost these two species. In so doing these areas have, *ipso facto*, also lost two critical functional groups,

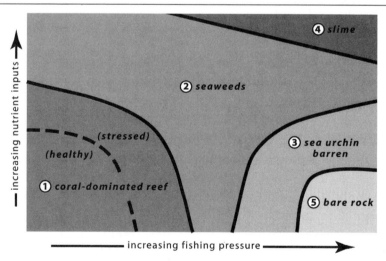

FIGURE 8 Possible Regime Shifts in Coral Reefs as Fishing Pressure and Nutrients Increase

As nutrients increase so does algal growth. As fishing increases the capacity for corals to regenerate decreases. Five different regimes are presented here.

(Adapted from Bellwood et al., 2004.)

BOX 5 Redundancy Is Not Necessarily a Dirty Word

Many years ago the phrase "being made redundant" was introduced in managerial circles as a euphemism for saying someone was being fired or retrenched. The phrase is a classic piece of efficiency-speak in which a worker's function is considered as already being performed by someone else and therefore that worker can be dispensed with. In discussions on ecosystem resilience, however, redundancy is not necessarily negative because the "workers" perform the same functions, but in different ways.

In the Caribbean, many functional groups that coral reefs depend upon have only a few species in them. The range of responses these functional groups can make to a disturbance is therefore limited. If each had more species, and each species had their own different response to a disturbance, reef response diversity would be enhanced.

The diversity of different functional groups of species—species that do different things in the ecosystem—contributes to a system's performance. Redundancy that increases an ecosystem's response diversity increases the resilience of its performance.

The coral reefs of the Caribbean had little redundancy to begin with. The impact of overfishing further reduced response diversity and now the region's most precious ecosystem has crossed a threshold into an undesirable regime. Had original diversity not been lost it is likely the reefs would have absorbed disturbances without a regime shift.

---

and two major shallow-water reef habitats: the elkhorn and staghorn zones (Bellwood et al. 2004).

Functional groups of fish are often classified by their position in the food chain (for example, the roles they play as predators and herbivores, referred to as trophic functional groups), and here again the Great Barrier Reef has more species of fish than the Caribbean in all trophic functional groups.

Consider, for example, the importance of three functional groups that play different and complementary roles in preconditioning reefs to permit the recovery of corals: bio-eroders, scrapers, and grazers. Bio-eroding fish remove dead corals, exposing the hard reef matrix for settlement of new corals. Scrapers directly remove algae and sediment by close cropping, thereby facilitating the growth and survival of coralline algae and corals. Grazers remove seaweed, reducing coral overgrowth by macro algae. Lose any of these functional groups and you diminish the capacity of the reef ecosystem to absorb disturbance, regenerate, and retain critical functions.

Compared with the Great Barrier Reef, the Caribbean had a relatively poor suite of species to begin with. Overfishing led to removal of entire functional groups causing the system to cross thresholds into new regimes in which fleshy macro-algae dominated. And without those functional groups it is impossible to return to the old regime of hard corals.

Some have suggested that engineering approaches might be applied to rebuilding coral reefs. Indeed, this is actually being attempted in the Indian Ocean following the destructive passage of the 2004 tsunami. The approach involves transplanting corals and constructing miniature artificial reefs. However, none of these interventions actually work at meaningful scales or provide realistic solutions to the increased global threats to coral reefs because they fail to reverse the root-causes of the regional-scale degradation (Adger et al. 2005). Realistically, it is only a restitution of the natural regeneration processes that will enable the wide spread return of hard corals. Consequently, restoration efforts should focus on improving water quality and restoring depleted fish stocks to enhance the innate resilience of coral reefs.

This is why ecologists are advocating that an important part of the response to this situation is a dramatic increase in the size and rate of establishment of "no take areas" where fishing and other human activities are prohibited, to enable the stressed populations of critical species of fish to regenerate.

However, as is being discovered, simply declaring an area protected doesn't necessarily protect anything if inadequate resources are provided to manage the area, or local people are not included in the conception or running of the area.

Coral Reefs Are Social-Ecological Systems

The Reefs at Risk project (Burke and Maidens 2004) evaluated the effectiveness of Marine Protected Areas (MPAs) in the Caribbean and found that there are 285 designated MPAs covering about 20 percent of the region's reefs. However, when they examined how these areas were operated they concluded that only 6 percent were effectively managed. Nearly half were rated as having an inadequate level of management, and therefore offered little protection to the resources they were designed to protect.

Common reasons for the failure of MPAs are the lack of long-term financial support from the local community, which can usually be traced to a lack of local involvement in planning and a failure to share financial or other benefits arising from protection. These failures reflect a traditional approach to natural resource management in which there is little acknowledgement of the linkages between the social and ecological domains of a system. And when it comes to the social domain of the Caribbean, it needs to be kept in mind that the Caribbean comprises an amazing diversity of countries and histories. More than any other case study in this book, the Caribbean faces formidable challenges in the nature of multiple jurisdictions and widespread poverty.

The history of the Caribbean as a whole has been one of colonial exploitation and resource extraction as a number of European powers maintained an often-brutal reign over various parts of the Caribbean (in some cases stretching back over five hundred years). When independence was ceded the infrastructure and assistance to help protect and develop resources was usually missing. They began their independent lives with low adaptability.

The management of the Caribbean, comprised of the territorial waters and coasts of thirty-five different countries and territories, is characterized by uncoordinated efforts. Contrast that to the situation of the Great Barrier Reef, controlled by one jurisdiction, the Australian government, with the capacity to invest considerable resources into managing a reef system that has not experienced the same degree of overexploitation seen in the Caribbean.

Resilience and the Coral Reefs of the Caribbean

The resilience of a social-ecological system depends on many factors, some of which are beyond the control of people living in a region. In the case of the Caribbean coral reefs, though the reef ecosystems had a good diversity of functional groups of organisms, they had very few species within functional groups (little response diversity) compared to other reef systems around the world, such as the Great Barrier Reef. This meant the system was more vulnerable to overfishing, pollution, urban runoff, and coastal development. Because of the low response diversity, whole functional groups have been effectively lost, and in many places hard coral no longer has the capacity to regenerate after a disturbance.

Even though the problems are largely understood, the Caribbean as a social-ecological system doesn't have much capacity to address the main issues that are threatening it. It has low adaptability.

4

In the Loop:

Phases, Cycles, and Scales—Adaptive

Cycles and How Systems Change

Thresholds as discussed in chapter 3 are relatively easy to appreciate because most people have experience of some aspect of an ecosystem that has gone bad and stayed that way. It might be a stinking lake that was once a popular fishing spot, or an eroded landscape that was once an agricultural breadbasket. It might be an entire swath of towns like the Rust Belt in the United States where industries are no longer competitive, or a vast area like the Aral Sea that has become a biological desert. These places have crossed a threshold into a regime in which the controls (feedbacks) are different, and it won't be easy to return to the way things were.

There is a much higher likelihood of crossing a threshold into a new regime if you are unaware of its existence. This can easily occur because resilience (which, as we discussed in chapter 3, can be defined as the distance to a threshold) is a multi-faceted measure of a social-ecological system that changes over time.

How do you identify and nurture resilience? Why do we frequently ignore it? Why do we allow something of value to degrade to a level that will impact human welfare?

This next building block of resilience thinking describes how systems move through different phases. It helps us better understand the dynamic nature of social-ecological systems and in so doing goes some way to answering these questions.

The Cycles of Life

You are born, you develop, you mature (maybe have children), and then you die. You move through different phases in the cycle of life. Families have cycles, businesses have cycles, nations and ecosystems have cycles. They are all around us, and we exist as part of cycles occurring over a range of scales in time and space. You might be surprised at the similarity between the various cycles at work around you.

One important aspect about cycles is recognizing that things happen in different ways according to the phase of the cycle the system happens to be in. Sometimes things change gradually, sometimes rapidly. Sometimes surprises are more likely, sometimes innovation has a greater chance of taking off.

By studying ecosystems all around the world, researchers have learned that most systems of nature usually proceed through recurring cycles consisting of four phases: rapid growth, conservation, release, and reorganization (Gunderson and Holling 2002). The manner in which the system behaves is different from one phase to the next with changes in the strength of the system's internal connections, its flexibility, and its resilience.

This cycle is known as an adaptive cycle (Gunderson and Holling 2002) as it describes how an ecosystem organizes itself and how it responds to a changing world. The notion of an adaptive cycle developed as a useful metaphor for describing change in ecological systems. However, it also has relevance for how social systems and social-ecological systems change through time. And though we use the term "cycle," the really important point is the existence of the four phases.

Although it is ecologists who have most thoroughly documented the adaptive cycle, an Austrian economist, Joseph Schumpeter, sparked the original idea. Schumpeter, who was influential in the first half of last century, analyzed the economy's boom and bust cycles, and described capitalism as a "perennial gale of creative destruction" (Schumpeter 1950). "Creative destruction" is a term now used to describe the disturbances that periodically punctuate the adaptive cycle. It breaks down stability and predictability but releases resources for innovation and reorganization.

Four Phases of the Adaptive Cycle

Understanding the significance of a system's internal connections, its capacity to respond to disturbance, and how these aspects change from phase to phase contributes to resilience thinking. This understanding is also important for policy and for managing natural resources because it suggests there are times in the cycle when there is greater leverage to change things, and other times when effecting change is really difficult (like when things are in gridlock). The kinds of policy and management interventions appropriate in one phase don't work in others.

The Rapid Growth Phase (or r Phase)

Early in the cycle, the system is engaged in a period of rapid growth as species or people (as in the case of a social system such as a new business venture) exploit new opportunities and available resources. r is the maximum rate of growth in growth models.

These species or actors (referred to as "r-strategists" in ecosystems), make use of available resources to exploit every possible ecological or social niche. The system's components are weakly interconnected and its internal state is weakly regulated.

The most successful r-strategists are able to prosper under high environmental variation and tend to operate over short timeframes. In ecosystems they are classically the weeds and early pioneers of the world (alder on newly exposed sites in northern forests, or dock and pigweed on cleared lands). In economic systems, they are the innovators and entrepreneurs who seize upon opportunity (think of the explosive growth of Google and other dot com companies). They are start-ups and producers of new products who capture shares in newly opened markets and initiate intense activity. At higher scales we can think of the emergence and rapid growth and expansion of new societies, and even nations.

The Conservation Phase (or K Phase)

The transition to the conservation phase proceeds incrementally. During this phase, energy gets stored and materials slowly accumulate. Connections between the actors increase, and some of the actors change, though by the end of the growth phase few, if any, new actors are able to establish.

The competitive edge shifts from opportunists (species, people, or organizations that adapt well to external variability and uncertainty) to specialists who reduce the impact of variability through their own mutually reinforcing relationships. These "K-strategists" (where K is the parameter for "carrying capacity" or maximum population size in growth models) live longer and are more conservative and efficient in their use of resources. They operate across larger spatial scales and over longer time periods. They are strong competitors.

In a growing business this often translates to a move toward more specialization and the greater efficiencies of large economies of scale: bigger machines, bigger outputs, smaller costs per unit, larger profits over longer timeframes (for example, a steelmaking business that grows from a local producer to a national and then a global company).

As the system's components become more strongly interconnected, its internal state becomes more strongly regulated. Prospective new entrants or new ways of doing things are excluded while capital grows (though it becomes increasingly harder to mobilize). Efficiency increases and the future seems ever more certain and determined.

In an ecosystem, the capital that accumulates is stored in resources such as biomass. Increasingly, more of it becomes bound up in unavailable forms, like the heartwood of trees and dead organic matter. An economic system's capital can take the form of built capital (machines, buildings) and human capital (managerial and marketing skills and accumulated knowledge).

The growth rate slows as connectedness increases, the system becomes more and more rigid, and resilience declines. The cost of efficiency is a loss in flexibility. Different ways of performing the same function (redundancy) are eliminated in favor of doing the function in just the most efficient way.

Increasing dependence on existing structures and processes renders the system increasingly vulnerable to disturbance. Such a system is increasingly stable—but over a decreasing range of conditions.

The Release Phase (or Omega Phase)

The transition from the conservation phase to the release phase can happen in a heartbeat. The longer the conservation phase persists the smaller the shock needed to end it. A disturbance that exceeds the system's resilience breaks apart its web of reinforcing interactions. The system comes undone. Resources that were tightly bound are now released as connections break and regulatory controls weaken. The loss of structure continues as linkages are broken, and natural, social, and economic capital leaks out the system.

In ecosystems, agents such as fires, drought, insect pests, and disease cause the release of accumulations of biomass and nutrients. In the economy, a new technology or a market shock can derail an entrenched industry. In each case, through the brief release phase, the dynamics are chaotic. But the destruction that ensues has a creative element. This is Schumpeter's "creative destruction." Tightly bound capital is released and becomes a source for reorganization and renewal.

The Reorganization Phase (or Alpha Phase)

In the chaotic release phase uncertainty rules; all options are open. It leads quickly into a phase of reorganization and renewal. Novelty can thrive. Small, chance events have the opportunity to powerfully shape the future. Invention, experimentation, and reassortment are the order of the day.

In ecosystems, pioneer species may appear from elsewhere, or from previously suppressed vegetation; buried seeds germinate; new species (including nonnative plants and animals) can invade the system. Novel combinations of species can generate new possibilities that are tested later.

In an economic or social system, new groups may appear and seize control of an organization. A handful of entrepreneurs released in an omega phase can meet and initiate a new renewal phase—turn a novel idea into a success (Nike shoes began in just this way). Skills, experience, and expertise lost by individual firms may coalesce around new opportunities. Novelty arises in the form of new inventions, creative ideas, and people.

In systems terms, the release phase is chaotic—there is no stable equilibrium, no attractor, no basin of attraction. The reorganization phase begins to sort out the players and to constrain the dynamics. The end of the reorganization phase and the beginning of the new rapid growth phase is marked by the appearance of a new attractor, a new "identity."

Early in renewal, the future is up for grabs. This phase of the cycle may lead to a simple repetition of the previous cycle, or the initiation of a novel pattern of accumulation, or it may precipitate a collapse into a degraded state (in social systems, a poverty trap).

Usually, a system passes through an adaptive cycle by moving through the four phases in the order described here (i.e., rapid growth to conservation to release to renewal). But this is not necessarily so. Systems cannot go directly from a release phase back to a conservation phase, but almost all other moves can occur.

Of Budworms and Social-Ecological Systems

A good example of adaptive cycles in ecosystems comes from the spruce/fir forests that grow across large areas of North America, from Manitoba to Nova Scotia and into northern New England. Among the forests' many inhabitants is the spruce budworm, a moth whose larvae eat the new green needles on coniferous trees. Every 40 to 120 years, populations of spruce budworm explode, killing off up to 80 percent of the spruce firs.

Naturally, resource managers looking after the forest wanted to control the damage caused by the budworms. However, their first efforts were carried out without an understanding of the cycle the forest was going through.

Following World War II, a campaign to control spruce budworm became one of the first huge efforts to regulate a natural resource using pesticide spraying. The aim was to minimize the economic consequences of the pest on the forest industry. Initially, it proved a very effective strategy, but like so many efforts in natural resource management that are based on optimizing production, it soon ran into problems.

In a young forest, leaf/needle density is low, and though budworms are eating leaves and growing in numbers, their predators (birds and other insects) are easily able to find them and keep them in check. As the forest matures and leaf density increases the budworms are harder to find and the predators' search efficiency drops until it eventually passes a threshold where the budworms break free of predator control, and an outbreak occurs.

While the moderate spraying regime avoided outbreaks of budworms, it allowed the whole forest to mature until all of it was in an outbreak mode. Outbreaks over a much greater area were only held in check by constant spraying (which was both expensive and spread the problem). The early success of this approach increased the industry's dependence on the spraying program, intensified logging and spawned the growth of more pulp mills.

Now there was a critical mass of tree foliage and budworms. The whole system was primed for a catastrophic explosion in pest numbers. The managers in this system were becoming locked into using ever-increasing amounts of pesticide because the industry wouldn't be able to cope with the shock of a massive pest outbreak. The industry had little resilience, and yet the continued use of chemicals was only making the problem worse. They had created a resource-management pathology.

The industry acknowledged the looming crisis and engaged ecologists to see how they might tackle the problem from a systems perspective. In 1973, a new analysis of the dynamics of the fir forests was presented by C. S. Holling, one based on the adaptive cycle.

Forest regions exist as a patchwork of various stages of development. The cycle for any one patch begins in the rapid growth phase, when the forest is young. The patch then proceeds through to maturity as described above, and eventually, following some 40 to 120 years of stable and predictable growth, the cycle tips into the release phase. The larvae outstrip the ability of the birds to control them, larvae numbers explode, and the majority of forest trees are killed. Their rapid demise opens up new opportunities for plants to grow, and during the reorganization phase the forest ecosystem begins to reestablish itself. The cycle then repeats.

With this understanding of the cycle and the key changing variables that drive the system, the forest managers were able to fundamentally modify the manner of their pest control. Rather than continually using low doses of pesticide over wide areas they switched to larger doses applied less frequently at strategic times over smaller areas. They reestablished a patchy pattern of forest areas in various stages of growth and development rather than keeping wide areas of forest primed for a pest outbreak.

The forest industry also changed through the process, moving to regional leadership with a greater awareness of the ecological cycles that underpinned the forest's productivity.

From Budworms to Resilience Thinking

The example of the spruce budworm and the fir forest is important on many levels as it was in part the genesis of what has become resilience thinking. During his investigations, C. S. "Buzz" Holling proposed that the key to sustainability was an ecosystem's capacity to recover after a disturbance. He also recognized that the ecosystem and the social system needed to be viewed together rather than analyzed independently, and that both went through cycles of adaptation to their changing environments. Adaptive cycles don't just happen in nature, they happen in communities, businesses, and nations.

His proposal catalyzed the thinking of ecologists and researchers (with an interest in systems) all over the world because similar patterns were being identified everywhere social-ecological systems were being studied. Over the decades since then the models and the thinking associated with managing for resilience have gone through much refinement but the two core areas remain at its heart: the fact that social-ecological systems can exist in different stable states and that they constantly move through adaptive cycles over many linked scales (which we discuss later in this chapter).

Fore Loops and Back Loops

Taken as a whole, the adaptive cycle has two opposing modes. A development loop (or "fore" loop), and a release and reorganization loop (or "back" loop) (see figures 9 and 10). The fore loop (sometimes called the front loop or forward loop) is characterized by the accumulation of capital, by stability and conservation, a mode that is essential for system (and therefore human) well-being to increase. The back loop is characterized

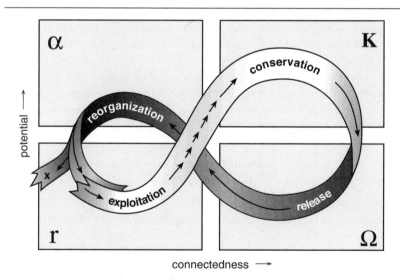

FIGURE 9 The First Version of the Adaptive Cycle

The first versions of the adaptive cycle pictured it as a figure 8 in two dimensions with the axes being connectedness and potential. Potential reflects accumulated growth and storage (biomass that is increasingly inactive like heartwood in trees or leaf litter). The use of the simpler loop, as shown in figure 10, has been adopted because it better reflects the passage from release to reorganization in some systems. However, because the adaptive cycle in the shape of the number 8 (as shown here in figure 9) was the original version it has iconic value, and it is often seen as a symbol of studies on resilience and adaptive cycles. *(From Gunderson and Holling, 2002.)*

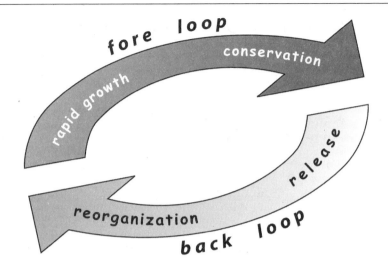

FIGURE 10 A Simple Representation of the Adaptive Cycle

The rapid growth and conservation phases are referred to as the fore loop with relatively predictable dynamics and in which there is a slow accumulation of capital and potential through stability and conservation. The release and reorganization phases are referred to as the back loop, characterized by uncertainty, novelty, and experimentation and during which there is a loss (leakage) of all forms of capital. The back loop is the time of greatest potential for the initiation of either destructive or creative change in the system.

by uncertainty, novelty, and experimentation. The back loop is the time of greatest potential for the initiation of either destructive or creative change in the system. It is the time when human actions—intentional and thoughtful, or spontaneous and reckless—can have the biggest impact.

It is important to reemphasize that the adaptive cycle is not an absolute; it is not a fixed cycle, and many variations exist in human and natural systems (see figure 11). A rapid growth phase usually proceeds into a conservation phase but it can also go directly into a release phase. A conservation phase usually moves at some point into a release phase but it can (through small perturbations) move back toward a growth phase. Clever managers (of ecosystems or of organizations) often engineer this in order to prevent a large collapse in the late conservation phase. That is, they avoid a release phase at the scale of concern (the whole forest or the organization) by generating release and reorganization phases at lower scales thereby preventing the development of a late K phase at the scale of concern.

Cycles of Nature and People

The easiest way to appreciate adaptive cycles is to observe them. Think of a forest, going through a succession from pioneer to climax species. This is the front loop that sees the forest resources slowly being accumulated and locked up in the trees and the various organisms they support. It's a long reasonably predictable phase of increasing growth. The longer it persists the more efficient it becomes in using the resources, and in so doing it eventually locks those resources up. As this occurs, the forest becomes less resilient, and more vulnerable to shocks and disturbances. At some point, inevitably, the forest will experience a disturbance such as a fire, storm, or pest outbreak big enough to precipitate a collapse, releasing accumulated nutrients and biomass. The longer the forest has been in the late conservation phase, the smaller is the size of the disturbance needed to send it into release. By comparison to the fore loop, the back loop is brief. The forest reorganizes, and sets up the start of a new cycle.

But think also of human systems. Consider a new business that builds houses. Being new, the business is trying out new and innovative ways of doing things, and is keen to build up its market. It proves successful and starts growing. Over time it starts adapting to its own

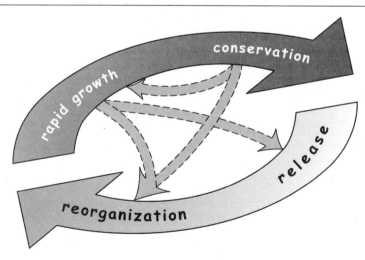

FIGURE 11   Variants of the Adaptive Cycle

Transitions are possible (and have been observed) between all phases except from the release or reorganization phases directly to the conservation phase.

success by being more efficient at doing the things that it does well. Resources are locked up in doing things in the most efficient manner such as in buying equipment for building houses in one set manner or tailoring the product to specifically meet the needs of just one subset of the market.

In so doing, however, the business is becoming less resilient as it concentrates on doing things in one or only a few different ways. If it invests in doing a wide variety of things, or in doing each thing in several different ways, it's being less efficient than focusing on just the things it does especially well. However, this loss of flexibility means it's more vulnerable to shocks and disturbances, such as an economic recession in which people cut back dramatically on house building or the appearance of a competitor who provides different kinds of houses. The business is totally dependent on houses being built in a particular way so it goes broke. It has entered the back loop. The resources it had locked up (people and capital) are now released and made available to the remnants of the company and new innovators that will be the trendsetters of the next cycle. The original company may or may not be a part of it.

Consider the global energy market and how it is dominated by fossil fuels. For many years scientists and economists everywhere have been talking about the need to move to alternate and renewable forms of energy production and yet we seem locked into our dependence on fossil fuel. For many it seems irrational, but it's in keeping with the adaptive cycle in which the front loop has experienced a slow, predictable pattern of growth in the fossil fuel industry in which resources are locked up into doing things in the most efficient way. But, in so doing, it has lost the capacity to do things in different ways. Too many institutions and businesses are now dependent on maintaining the status quo of fossil fuel usage. Innovation for alternatives is either quashed or receives inadequate support. A consequence of this is that it lacks resilience. Fossil fuels have a limited life span and at some point a shock will lead to a collapse, a major disruption of the industry. Whether or not this is a sudden, costly, painful change will depend on whether the industry has the ability to transform itself ahead of time.

If it's a sudden collapse, in the back loop that follows, the resources it has locked up will be released toward innovation in different areas. The later this happens, the bigger will be the losses in the release phase.

And the next forward loop of the cycle will be different from that which would take place if the reorganization were to occur earlier. We can't say exactly how it will be different, but if the collapse phase is bigger it's likely that at least the start of the next loop will witness a period of reduced human well-being.

There is, of course, the option to choose a transition to an energy mix that is sustainable (as is being attempted in some European countries). Such actions and induced changes constitute either a reverse move, from late conservation back to early conservation phase through small-scale changes, or a direct move at the scale of the whole industry through a rapid renewal phase to a new front loop, attempting to minimize the costs of the release phase.

Because the back loop is a time of uncertainty and big change, in which the usual order undergoes significant and unpredictable rearrangement, it is feared and held off by those in power. However, no system can stay in, or be kept in, a late conservation phase indefinitely. Unless there is a deliberate effort to simplify the complexity, to release some of the potential and slide back toward the rapid growth phase, or (as is suggested) engineer a very rapid, minimal cost conservation-to-reorganization transition, a significant back loop of one form or another is inevitable.

Dangers of the Late *K* Phase

We stress that the fore loop is crucial for capital accumulation (all forms of capital). It is where levels of human well-being can be raised. Capital does not accumulate during periods of release and renewal. From rapid growth to mid-conservation is a period of accumulating capital but also a time of high to moderate resilience.

However, things are not so positive as you move into the late *K* (conservation) phase where the system begins to become locked up. This situation is characterized by:

- Increases in efficiency being achieved through the removal of apparent redundancies (one-size-fits-all solutions are increasingly the order of the day);
- Subsidies being introduced are almost always to help people *not* to change (rather than *to* change);
- More "sunk costs" effects in which we put more of our effort into

BOX 6 Of Levees, Cycles, and Windows

The adaptive cycle explains why sometimes a good idea by itself isn't enough to cause something to happen; it needs a window of opportunity to really work. While it may take off during the reorganization phase, if left till the K phase you've probably missed your opportunity.

Take, for example, an innovative proposal in the lower Goulburn-Broken Catchment to dismantle a series of flood control levees to reestablish the natural function of a floodplain adjoining the Goulburn River. The flood levees were established around 100 years ago following a series of damaging floods. As with most of the examples of command and control presented in this book, the flood levees solved one short-term problem but created a series of larger scale issues. In this case, small and medium floods were kept off the extensive River Red Gum flood plains to the north. This protected the crops being grown there but removed the regular flooding that sustained the River Red Gum ecosystems. Instead of allowing the floodwaters to cover the flood plains, the floodwaters were forced down the main channel causing extensive erosion and environmental damage to the river channel.

While the levees contained small-to-medium floods, larger floods (which occurred most decades) were not contained. Not only did they breach the levees on the northern side, pouring out over the flood plain, they also breached the levees on the southern side, allowing floodwaters to damage valuable irrigation land.

It was very much a lose-lose situation. The levees cost a lot to maintain, harmed a valuable ecosystem, failed about once every 10 years, and exacerbated the damage caused by moderate floods. Their only value was to protect low-value cropping land from small floods. It is interesting to note how we can stick with bad ideas over long periods because innovation is so challenging to introduce.

The spring floods of 1993 were particularly bad causing AUS$1 million damage to the levees and AUS$20 million in losses and damage to adjoining infrastructure and agricultural production.

The scale of damage finally galvanized action. The Catchment Management Authority proposed to dismantle the levees on the northern side of the river. This would allow floodwaters to once again pour over some of the original floodplain. The River Red Gum floodplain would be rejuvenated, the main river channel would not be scoured in times of flood and the valuable agricultural lands to the south of the river would be significantly better protected.

The main cost of the scheme would be the purchase of 10,000 hectares of privately owned land on the floodplain by the government, the cost of which would be less than the damage bill of the single event in 1993.

Unfortunately, the proposal, which was touted by most parties as an environmental triumph with economic benefits, took several years to work up and was met with resistance from a few people with a vested interest in keeping parts of the floodplain in private ownership. More studies were called for to clarify minor aspects of the proposal and the process dragged on. The further away from the catalyzing floods of 1993, the fewer people seemed to care whether or not it got up.

In 2005, following the worst drought on record, the window of opportunity appears to have closed on this particular proposal, at least until the next flooding event, when this region may enter another phase of creative destruction (in its passage through the adaptive cycle). Then, with the proposal all ready, it may have its window to fly.

And, if you think about it, natural flooding in this system really is creative in its destruction. It destroys weeds; moves debris around; deposits silt and nutrients; triggers breeding by fish, amphibians, and water birds; and replenishes billabongs (bodies of water like small oxbow lakes), wetlands, and streams. The levee system was an attempt to control creative destruction. Unfortunately, it got rid of the creative and simply left the destruction.

---

continuing with existing investments rather than exploring new ones (the Concorde effect);
- Increased command and control (less and less flexibility);
- A preoccupation with process (more and more rules, more time and effort devoted to sticking to procedures);
- Novelty being suppressed, with less support for experimentation; and
- Rising transaction costs in getting things done.

Capital doesn't accrue in the late K phase either, and the likelihood of a major collapse is high. So, if the system is in late K, the first question is how to undo some of the constraints. Any release phase is costly and unpleasant and involves loss of capital (social, economic, and natural), so if a release seems inevitable, then the question becomes: How can we navigate a graceful passage through the back loop?

While a system may stay in a late conservation phase for some time, it can't stay there forever; complex systems don't work that way. The costs involved in staying in this late conservation phase increase over time. At the scale of a society or a nation, when those costs exceed the benefits of all the fix-up solutions, the society collapses (Tainter 1988).

A Window of Opportunity

As should be evident by now, a back loop is not all bad. It is a time of renewal and rejuvenation, a period of new beginnings and new possibilities—hence its description as a period of creative destruction.

Droughts and floods can trigger creative destruction but so too can economic recessions, wars, and the deaths of powerful leaders (consider the discussion on adopting new ways for thinking and paradigm shifts in chapter 2). These events may be traumatic and destroy property and people but they also make it possible for new beginnings. And those new beginnings can often grow to be ruling paradigms in the next front loop. They are critical times to achieve change and reform in a constantly moving social-ecological system.

Think back to the Everglades. Its history has been characterized by cycles of disturbance, release, reorganization, and incremental growth. It is currently in a state of legislative gridlock in which it may take a new disturbance to move things along.

Once you move out of the back loop, opportunities for innovation and novelty shrink and eventually shut down as the slow grind of the front loop starts to squeeze out new entrants, new ideas, and different ways of doing things.

The Importance of Scale(s)

When we talk about adaptive cycles in social-ecological systems it is easy to become too focused on the specific scale in which we're interested. If we're talking about a farm or a business, we usually think only of that farm or that business. But the scale in which we are interested is connected to and affected by what's happening at the scales above and below, both in time and space. At each scale the system is progressing through its own adaptive cycle, and the linkages across scales play a major role in determining how the system at another (linked) scale is behaving. We've already described how generating disturbances at lower scales can keep a system at a higher scale from progressing to a late K phase.

What all this means is that any system you can imagine is actually composed of a hierarchy of linked adaptive cycles operating at different scales (both in time and space). The structure and dynamics of the system at each scale is driven by a small set of key processes and, in turn, it is this linked set of hierarchies that govern the behavior of the whole system. This linked set of hierarchies is referred to as a "panarchy."

The term "panarchy" was originally coined by Buzz Holling and Lance Gunderson to describe the cross-scale and dynamic character of interactions between human and natural systems. It draws on the Greek god Pan, a symbol of universal nature, to capture an image of unpredictable change, and fuses this with notions of hierarchies—cross-scale structures in natural and human systems. The term embodies notions that sustain the self-structuring capacity of systems (system integrity), allow adaptive evolution, and at times succumb to the gales of change.

It's important to get this multi-scale structure and dynamics clearly in mind, and it helps to consider a couple of examples. Think again about the conifer forests in North America. The finest scale (for our purposes) is the leaf, or individual conifer needle. The next scale up is the crown of the tree. Figure 12 shows the full range of space and time scales, for both the forest and for the processes that produce this structure.

Structure at the leaf scale, measured in centimeters, is driven by plant physiology and environmental conditions operating over time scales up to

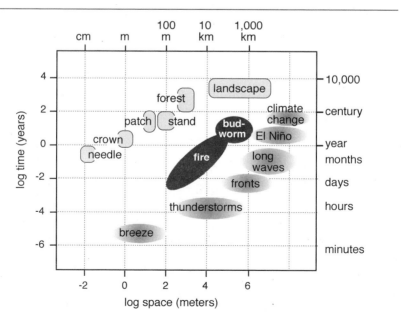

FIGURE 12 Time and Space Scales of the Boreal Forest
(From Gunderson and Holling, 2002.)

a year—the generation time of a needle. Canopies operate in scales of meters and cycle with a generation time of around ten years. Trees cycle with a generation time of a hundred years or more. At the forest patch structure the relative numbers and sizes of different tree species—working on scales of tens of meters—is driven by competition for light, water, and nutrients, and cycles over centuries. The structure of the whole forest operates at distance scales of kilometers, timescales of centuries. It is driven by processes such as fires, storms, and insect outbreaks. At the level of the landscape, and at scales of thousands of kilometers, climate, geomorphologic, and biological processes drive changes at time scales from centuries to millennia.

Interacting with the natural system scales is a set of social scales of human activities using and working in and around the forest. There are time and space scales relating to individual workers, their families, and their communities. There are also time and space scales relating to the operation of the forestry business, the equipment it uses, the industry it's a part of, and the overall demand for wood products that relate to bigger cycles of economic prosperity and growth. The cross-scale dynamics of the natural and social components of this complex system constitute the panarchy that is an interlinked system of people and nature.

Very importantly, the processes that produce these panarchy patterns are in turn reinforced by those patterns—that is, *the patterns and processes are self-organizing*. This is a key aspect of complex adaptive systems (see chapter 2).

Connecting Across Scales

Of particular interest are linkages across scales. They are a key aspect of the multiscale adaptive cycles that make up a panarchy. What happens at one scale can influence or even drive what's happening at other scales.

Ignoring the effects of one scale on another (cross-scale effects) is one of the most common reasons for failures in natural resource management systems—particularly those aimed at optimizing production. The lesson is that you cannot understand or successfully manage a system—any system, but especially a social-ecological system—by focusing on only one scale. So often people concentrate solely on the scale of direct interest to them (their farm, their company, their catchment, or their country), but the structure and the dynamics at that scale, and how the system can and will respond at

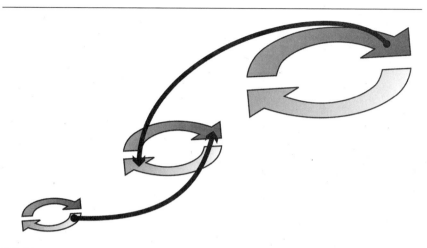

FIGURE 13 Panarchy Refers to Hierarchies of Linked Adaptive Cycles

that scale, strongly depends on the states and dynamics of the system at the scales above and below.

Some obvious examples: individual farmers who wish to clear trees from their lands, or drain excess irrigation water into rivers, may be constrained by higher level regulations. Their ability to sell their produce may be affected by changes in preferences at the scale of the market. And, in times of severe hardship (a drought) the help they may or may not get from the scale of organization above (often some level of government), depends on the state the system happens to be in, at that scale, at that time. For example, if a nation's economy is strong going into a drought there might be strong community demand for the government to provide drought assistance to farmers. However, if a nation is in economic depression then drought assistance is unlikely.

The recovery path of a forest patch that has been devastated by a fire or cyclone depends on the availability of seeds of the many species in the surrounding mature forest. That is, it depends on the "memory" of the system at the scale of the whole forest. The recovery pattern of communities of people that have been subject to a devastating shock (environmental, economic, or social) will depend on the memory of how to respond, embedded at the higher scale of the society in which the community exists.

Bottom-up linkages can be equally important. At the scale of a forest the individual patches, going through their own (faster) cycles, may produce a kaleidoscopic effect in which the forest, as a whole, may appear to remain much the same if considered as an average of the patches. But if many of the patches get into the same phase of their cycle (too much synchronization) a change in one can trigger others and the effect can cascade upward to a change in the whole forest. This is what happened when pesticides were used to control the spruce bud-worm. The pesticide use prevented forest patches from succumbing to pest outbreaks but synchronized wider areas of the forest patches.

Continuing the fossil fuel example, if the fossil fuel industry were to collapse it would likely drive many economies into recession or depression. This economic downturn might tip many businesses, which operate on smaller scales (and which were not thinking about the fossil fuel industry scale), into their own back loops.

Sometimes the connection with larger scales can perpetuate a conservation (K) phase at smaller scales that might otherwise have ended. In the Everglades, the continued drive to develop, drain, and regulate the waters of the marshlands may have been abandoned if there hadn't been massive injections of resources from the federal government. Subsidies from a higher scale frequently prolong a K phase that would otherwise have moved into a back loop.

In the Goulburn-Broken Catchment (GBC), one of the important factors that allowed the region to create new levels of organization to address the crisis of the underground flood of the 1970s was the higher scales of state and federal government relaxing their control and allowing for the devolution of authority to local authorities. The success of the Goulburn-Broken Catchment Management Authority in addressing aspects of the groundwater problem served as an example to other regions and states that then established their own catchment management authorities. An innovation in one catchment propagated up to influence the larger scale.

But there's another interesting cross-scale effect in the case study on the GBC. In the 1970s, just as high rainfall caused an enormous rise in groundwater killing many of the region's fruit trees, England had just joined the European Union and Australia lost an important export market for its fruit. The price being paid for fruit from the GBC plummeted, and this served to mask the impact of the rising groundwater because now the produce from the dying trees wasn't worth much. Farmers were pulling out trees anyway. Had there not been this market impact it's believed a lot more would have been done at that time to address the problem of the rising groundwater. Another possibility may have been that the fruit industry in the GBC would have moved to another area.

Thresholds and Adaptive Cycles

Thresholds (chapter 3) and the adaptive cycle metaphor are both central to resilience thinking. Adaptive cycles describe how many systems behave over time, and how resilience varies according to the phase where the system lies. Thresholds represent transitions between alternate regimes. While the two concepts can sometimes be related in the pattern of a particular system's dynamics, this not always the case. They are different models used for different purposes, and it is not always possible to equate the dynamics of a basin of attraction with the dynamics of an adaptive cycle. Where they do coincide, however, alternate regimes generally represent a new adaptive cycle, indicating that the system has new structures and feedbacks.

When a social-ecological system crosses a threshold into an alternate basin of attraction (say a catchment becomes salinized or a coral reef stops regenerating) it often occurs during a back loop. Linkages that bound the system in the conservation (K) phase are broken during the release phase. Some species may fail to recolonize during the subsequent reorganization phase with the result that a loss of species occurs, possibly also a collapse in productivity. These changes may represent a new system regime, and a new adaptive cycle.

However, crossing a threshold may occur in situations where the adaptive cycle phase of disturbance is small relative to larger environmental changes. A patch of rangeland, for example, that shifts from a grassy regime to a shrubby regime doesn't mark a collapse in the system, at that scale. There is no unraveling or chaotic dynamics. If enough patches flip, however, it can trigger a collapse of the farm as an enterprise.

One example that illustrates how resilience, thresholds, and adaptive cycles are related is the successional changes that occur in the transitions from a beaver pond to a bog to a forest. Each of the stages in this well-known development sequence is distinctive in the way it looks and functions. Each marks a transformative change from the preceding system. The changes are transformative because in each case the system has a new way of functioning and new identifying variables come in to the system—bog plants in the first transformation, trees in the second. These are not alternate regimes in the same system; they are different systems. But each transformation occurs without a release phase—there is no catastrophic loss of nutrients, for example. This is an example of adaptive cycles involving progression from a rapid growth phase to a conservation phase and then briefly through a reorganization phase (as the new species enter) and then to a newly self-structuring growth phase of bog or (in the next round) forest. This type of development may be characterized as a series of steps, with each system a different step.

Each transition is a consequence of a loss of resilience in the existing system. Starting with the pond, as it gets shallower its ability to resist the invasion of bog plants gets smaller. Eventually it is overwhelmed during some small shift in pond depth (a small drought or seasonal effect) and the bog gets established. The bog grows vigorously until it reaches an advanced stage with lowering water levels and with reduced vigor and self-organizing ability. If, before this time, a fire should occur the likely consequence is a short back loop of release and reorganization, and a new invigorated bog growth phase. As sediments increase, however, and bog conditions diminish, trees gradually establish in the bog, first on hummocks, until a drought or some other disturbance gives the trees the upper hand throughout. The more developed the bog, the smaller the shock until tree invasion inevitably occurs—with no release—and reorganization into a forest system follows. Each transition is a consequence of a loss of resilience due to a slowly changing variable (water depth/saturated soil depth) with a threshold where feedbacks change and a regime shift (in this case a transformational one) takes place.

Complex Science and Back Loops

We've pointed out that the fore loop is generally slow compared to the back loop. In fact, most systems spend by far the majority of their time in a fore loop. So if you look around, by far the majority of what you see will be systems in fore loop phases.

It is therefore not surprising that nearly all our research, and all our management and policy development, has been done in and created for fore loop behavior. Almost none has been done in systems in their brief but critically important back loop periods.

Recognition of the enormous importance of the back loop is very recent. It is part and parcel of appreciating that we live a complex world. It is a major component of resilience thinking and an important point of difference with traditional science that has modeled the world based on the assumption that change is incremental and predictable. Yes, change *is* incremental and predictable in the fore loop but to limit our management and policy to the parts of the world that we understand while ignoring those parts that are difficult is to set ourselves up for failure.

Key Points on Resilience Thinking

- Social-ecological systems are always changing and many changes reflect a progression through linked adaptive cycles, on different scales of time and space, with each cycle consisting of four phases: rapid growth (r), conservation (K), release (omega) and reorganization (alpha).
- Most of the time, social-ecological systems are changing along the growth to conservation phases (fore loop) of the cycle in which growth and development are incremental, life is fairly predictable, and resources get locked up in doing things in an increasingly efficient manner. Optimization for immediate benefits can work in these phases (for a while).
- Inevitably, the conservation phase will end. The longer the conservation phase persists, the smaller the shock needed to end it and initiate a release phase in which linkages are broken and natural, social, and economic capital leaks out of the system. The system then reorganizes itself. In this back loop passage through release and reorganization, uncertainty and instability are high and optimization does not work.
- While the cycle just described is the most common pattern of system dynamics, other transitions between the four phases can, and do, occur.
- Linkages across scales are very important for how the system as a whole operates.
- By understanding adaptive cycles you gain insight into how and why a system changes; develop a capacity to manage for a system's resilience; and, most importantly, learn where and when various kinds of management interventions will, and will not, work.

CASE STUDY 4

Scenarios on the Lakes:

The Northern Highlands Lake District, Wisconsin

I f you were looking for a nice place to escape the rat race, you could do worse than buy a plot of land in the Northern Highlands Lake District (NHLD) in Wisconsin. It offers a landscape of diverse lakes, rich forests, and a range of recreational activities. However, if that's where you want to head then you'd better move fast because most of the prime locations are gone, and what's left is skyrocketing in price. Why? Because it's an area a lot of people want to move into. Unfortunately, as more people move into the area, things change and the future becomes increasingly uncertain.

It's a recurring story all around the world. Landscapes high in natural beauty are experiencing growing populations, increasing pressures on ecosystem services, environmental degradation, and falling amenity. Some describe it as "being loved to death," some as "killing the goose that laid the golden egg," and others as "environmental vandalism." The people who are already there want the extra resources that usually come with a growing population (which usually includes social and economic infrastructure) but bemoan the loss of their beloved environment as it existed in "the good old days." And new arrivals often become upset as the various values of the region decline—values which had them moving into the area in the first place.

In general, the slow erosion of the natural values of an area is accepted fatalistically, but sometimes there comes a point when the things that once made an area a nice place to visit, holiday, or invest in, seem to vanish. And when that happens, call it a tipping point or crossing a threshold, suddenly no one wants to be there, and the region begins to slide.

How do you make decisions that will avoid potential risks while taking advantage of potential opportunities? How might the NHLD plan for an uncertain future?

The NHLD in a Nutshell

The NHLD lies in the north of the state of Wisconsin. It contains around 7,500 natural lakes that together in area comprise over 13 percent of the region. Some four fifths of the region is forested. Lake Superior, the world's largest freshwater lake, lies a short distance to the north. The climate is heavily influenced by its proximity, giving cool summers and cold winters.

The region has experienced several periods of glaciation which have left a relatively flat landscape. When the last glaciers retreated twelve thousand years ago many lakes were formed. Unsurprisingly, lakes are the NHLD's most conspicuous and distinctive feature. Some occur in hollows in outwash gravel plains; others are formed in depressions in the ground moraine or were created by the melting of buried ice chunks.

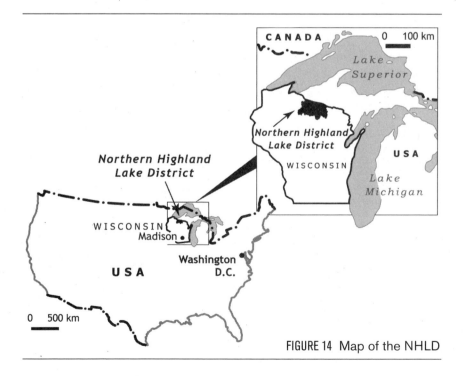

FIGURE 14 Map of the NHLD

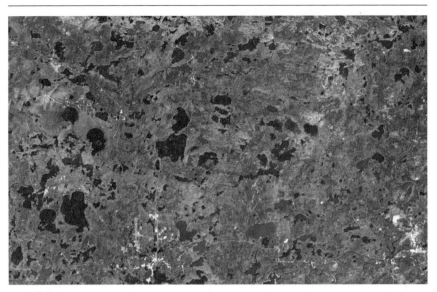

IMAGE 10

A satellite view of the lakes district in the northern highlands of Wisconsin. The lakes range from small ponds to vast expanses of water. *(Courtesy of the University of Wisconsin Environmental Remote Sensing Center.)*

They range in size from small, temporary ponds and darkly stained "bog" lakes to large expanses of water covering well over a thousand hectares. Depths range from one to more than thirty meters.

The NHLD has been sparsely inhabited for thousands of years. In the seventeenth century, European fur traders entered the region and transformed the lives of the Native Americans, its original inhabitants. In the nineteenth century, the expansion of the United States led the First Nations to surrender their sovereignty in exchange for land use, hunting, and fishing rights in a series of treaties.

In 1900, the population of the area was approximately twelve thousand. Over the past century it has grown to around sixty-five thousand permanent residents, with growth accelerating over the last three decades. Recreation and tourism are major components of the economy, and there has been substantial development of vacation and retirement homes around local lakes. Fishing is one of the major tourist attractions of the region. The region lies within a day's drive of several major urban centers, including Chicago, Milwaukee, and Minneapolis-Saint Paul.

The Crowding of the NHLD

Everyone has noticed it but no one has been too sure what it all added up to. The Lake District is simply not the place it once was. There are more people, more conflicts, and more tension.

In the decade leading up to the year 2000 the population grew by 15 percent and property values doubled. Highways connecting the region to the big cities have expanded, bringing more traffic and more visitors. The NHLD's urban centers are larger. International or national chains are more prominent in the business community, displacing many of the "old," traditional businesses (such as lodges and restaurants).

Almost all the lake shore that can be developed has been developed, and now the region is experiencing significant levels of redevelopment as older and smaller cottages are replaced with newer and larger houses. Development has led to the deforestation of riparian areas, invasion of exotic species, and the pollution of lakes through runoff and leaky septic tanks. In many areas the lake edges have also seen the removal of reed beds and woody debris, both being valuable habitat for wildlife. Fishing has thinned the larger fish on most public-access lakes. Anglers are common on the lakes but so too are highly polluting speedboats and personal watercraft.

Invasive species are a growing problem for both terrestrial and aquatic ecosystems. Lakes are being invaded by exotic fish (e.g., rainbow smelt), plants (e.g., purple loosestrife), and invertebrates (e.g., rusty crayfish). Many exotic species gain access to previously uninvaded lakes in bait buckets or by attaching to boats.

However, it's not all bad news. The increased population has seen improvements in health care. The development of service-orientated industries has increased, including niche businesses such as gourmet coffee shops and delicatessens that would normally only be found in bigger cities.

A History of Change

The NHLD is in rapid transition but, as with most regions, transitions have happened before. Twelve thousand years ago the current lake landscape was molded as the glaciers receded. Archaeological evidence suggests that humans moved in soon thereafter. This has been dubbed the Age of Discovery, and it was marked by the early disappearance of mega fauna such as the woolly mammoths.

Europeans entered the region about five hundred years ago through

IMAGE 11

An aerial view of the lakes landscape. *Photo credit: ©Steve Carpenter*.

the fur trade. Following their arrival, native populations decreased significantly while newcomers took ownership of the land.

The old-growth forest of the NHLD was extensively logged in the late nineteenth and early twentieth centuries to provide timber for the growing cities south of the NHLD. By the 1930s, much of the Northern Highlands had been clear-cut. Over four-fifths of the region has since been reforested.

The region is now entering a new age, in which woolly mammoths have been replaced by mammoth homes. Fishing and canoeing are competing with motor sports such as power boating and bush bashing in all-terrain vehicles.

Invasive species and emergent diseases are threatening the quality of the environment in the region. The quality of the fishing is declining. Warmer, shorter winters with less snow are impacting on the tourism industry as well as the region's ecosystems.

What do all these changes add up to?

Exploring Uncertain Futures

A traditional approach to exploring the future has been to extrapolate from the past. You look at past trends, construct a model that simulates how things are working and wind it forward. But experience has shown that there's limited value in this approach when attempting to explore the future of social-ecological systems. They are simply too complex, while the models are invariably too narrow in their focus. Plus, the future has a habit of throwing up surprises, a product of the complex nature of social-ecological systems, which often make the simulation model irrelevant.

Because many changes in the NHLD are happening simultaneously it's difficult to know what the cumulative impact will be. With so many changes happening at once, it seems hard to think about the future in a cohesive way. The problem facing those in charge is how to make the

---

BOX 7 One Size Does Not Fit All

Lake districts such as the NHLD are often managed as if the lakes are independent, similar, and only affected by factors local to a particular lake. Consequently, when it comes to setting up rules and regulations to manage activities such as sports fishing a one-size-fits-all approach is often applied. Careful analysis of such approaches, based on the NHLD, however has shown that that this approach simply doesn't work (Carpenter and Brock 2004). Such policies actually lead to the domino-like collapse of fisheries across the system of lakes.

Carpenter and Brock used detailed analyses of both fish population dynamics in a lake, and of angler movements and fishing efforts between the lakes. Because lakes differ significantly in terms of their attributes that determine fish population resilience (shoreline habitat is particularly important), a regulation that ensures all lakes will always be safe leads to big losses in angler benefit (fish harvest), and incentives for anglers to break the regulation (a decline in social resilience). If regulations are too lax, ecological resilience declines and some lakes collapse, leading to increased pressure on the others.

In either case, the collapse of fish populations on some lakes causes anglers to shift to other lakes creating a cascade of collapses and the degradation of fisheries in most lakes. Under one-size-fits-all management the natural resources of the entire landscape become more vulnerable to unwanted change.

The basic problem with one-size-fits-all is that it doesn't acknowledge that

- The lakes are inherently different
- There are linkages connecting the lakes

Applying the same regulation to all the lakes is bound to fail through loss of either ecological resilience or social resilience.

system that people like and want resilient in the face of future surprises. Rather than simulate the future, researchers from the University of Wisconsin decided instead to explore what the next twenty-five years might have in store for the NHLD by constructing a series of scenarios (Peterson et al. 2003a; Carpenter, in press).

Scenarios are not predictions of what *will* happen. They are an exploration of what *might* happen. They are structured narratives about the possible future paths of a social-ecological system (Peterson et al. 2003b). Rather than forecasting the future, they involve a group of experts working together with a representative cross section of local residents to explore what might happen to the region if certain trends are followed.

Scenario planning began after World War II as a type of war game analysis. Later it was used as a part of business planning. The oil company Royal Dutch Shell pioneered it in industry planning and, indeed, it played an important role in that company's success during and after the world oil price crises in the 1970s. More recently, scenarios have been used in global environmental assessments such as the Millennium Ecosystem Assessment. Scenario planning has also been used during the transition to democracy in South Africa, and in community planning exercises in the United States, Europe, and Asia.

The process of building a scenario involves bringing together the best available information on the current condition of the region (biophysical, social, and economic) and then identifying key uncertainties, vulnerabilities, sources of resilience, and the hopes and fears of people for the future of the region. These considerations lead to many scenarios. Typically the scenarios can be clustered into a few stories that capture the main points. In the case of the NHLD, four scenarios emerged (Carpenter in press).

The stories that are developed through this process about how the region might change over time usually identify a range of issues for ongoing discussion. The scenarios, therefore, are not so much the end of the process as the beginning of an ongoing debate and discussion.

Scenarios help organize information, and they are easy to understand. Scenario planning is also a good way to open discussion among different groups of people who might not otherwise interact. Scenarios allow us to consider several possible futures instead of trying to predict a single one. These possible futures are not likely to come true exactly as described in the scenarios, but they let us think in broader terms about the impacts of the plans and choices we make, and how to make the kinds of system regimes we might like more resilient.

Involving and engaging the local residents is an essential part of the overall process. Not only do they possess many insights on what drives a region, but scenarios can also help the people of the NHLD consider how they might prepare for possible change. It also encourages people to see their region as a social-ecological system—as a complex adaptive system in which no one is in control and which has the capacity to cross thresholds into an undesirable regime. Building scenarios through social networks helps people appreciate many aspects of resilience thinking.

And the process in many ways is just as important to building resilience as the scenarios it produces. Through people sharing and building social networks that span different areas and scale of operation, the community is in essence building trust and social capital that is basic to enhancing adaptability and resilience.

Scenarios for the NHLD

Four different scenarios were developed for the NHLD through a series of workshops in 2003 (Peterson et al. 2003a; Carpenter in press) involving

IMAGE 12

Networking by the lake: members of the NHLD community discuss scenarios of what the future holds for their region (seen here listening to Dr. Steve Carpenter). (*Courtesy of Susan Carpenter.*)

people from the NHLD and the University of Wisconsin. The NHLD people included officials from the county and Wisconsin Department of Natural Resources, members of area lake associations and the area's Native American tribes, local realtors and business owners, and part- and full-time residents.

Each scenario grows out of a shared baseline story that encapsulates what is known about the present and what is expected in the near future. Each scenario then traces a different sequence of events until 2027. While there is insufficient space here to describe each one in detail, the following descriptions provide a flavor of what was created. (See the 'Future of the Lakes' website for more details, http://lakefutures.wisc.edu).

The Common Baseline

Native Americans are an important component of the region's future development. On the Lac du Flambeau reservation the casino operated by the Lac du Flambeau Indians* is bringing in new wealth. The number of tribal residents has increased as members return to participate in the growing economy, and young people stay because jobs are available. The number of pupils in the Lac du Flambeau Public School has increased even as the school-age population in most of the NHLD has declined. Living resources—fish and game—on tribal lands are flourishing.

In the rest of the NHLD, things aren't so positive. The place is filling up. There are confrontations in county boards over land use and shoreline regulations. Communication is poor between nonresident lakeshore property owners and local residents. Many residents are unhappy about the replacement of old businesses—lodges, restaurants, and stores—by outside chains.

Over the years, the environment has been changing. Weather is seemingly more variable. Warm, wet winters reduce skiing and snowmobiling. There is debate about whether all-terrain vehicles could or should fill the economic niche once filled by snowmobiles. Conflicts arise between jet ski lovers and haters and there are disagreements about how much of the landscape should be devoted to loud, motorized activities versus quiet, muscle-powered recreation.

---

*The Lac du Flambeau area of the NHLD has been inhabited by the same band of native Indians since 1745 when Chief Keeshkemun led the band to the area. The band acquired the name Lac du Flambeau—"Lake of Torches"—from its practice of harvesting fish at night by torchlight. The Lac du Flambeau reservation has 260 lakes, sixty-five miles of streams, lakes, and rivers and twenty-four thousand acres of wetlands.

Anaheim North

In the first scenario, called "Anaheim North" (and also known as Wal-Mart nation), tourism takes over. Theme parks, big businesses, and urban sprawl cover much of the landscape (as can be seen in the city of Anaheim in California). Population and commercial activity increase, but many of the jobs in the NHLD are low paying and much of the profit of the theme parks does not stay in the NHLD. Locally owned businesses become less common. Problems with urban sprawl and pollution intensify. The dramatic increase in annual visitors also leads to an expansion of the Lac du Flambeau Casino.

Motorized recreation replaces muscle-powered recreation, except in the most remote areas and on private tracts of land. Public hunting and fishing lands are heavily harvested, and quality hunting and fishing experiences are found only in a few remote sites and on large private landholdings.

By 2027, the region is transformed. The population has almost doubled, the economy is larger, and so are the size of businesses and the role of corporations based outside the NHLD. Much of the profit generated from tourism flows out of the NHLD. The gap between rich and poor has grown, urban sprawl is notable around the region's main centers, and air, water, light, and noise pollution are increasingly common problems. The level of trust and cooperation among people in the region has declined to resemble that of other urban regions across the United States.

Walleye Commons

The second scenario presents a different future. In this scenario the driving force of change is deregulation. The state government, crippled by financial crisis, relaxes shoreline management practices and building restrictions. Along with difficult-to-control ecological disturbances, such as the spread of chronic wasting disease in deer and invasive species, the landscape changes, encouraging many tourists and residents to pack up and head to new destinations.

The visitor population declines as a result of intensifying conflicts over resource use, environmental deterioration, and collapse of a real-estate bubble. Despite economic hardship, the Lac du Flambeau tribe persists. Ecosystems recover slowly. The economy is smaller in 2027 than in 2002, but more diverse with contributions from ethno-tourism and slow recovery of recreational opportunities on feral ecosystems.

At the same time, the Lac du Flambeau tribe expands its land holdings, introduces experimental management strategies and experiences a cultural renaissance. Through their efforts, the quality of the lake and land slowly recovers. The name "Walleye Commons" refers to shared use of an ecosystem dominated by walleye, a favored game fish of the region.

Although the economy is much smaller that it was, many residents feel that their rural lifestyle and the gradually improving environment of the NHLD more than compensates for their low incomes.

IMAGES 13 & 14

Will the future of the NHLD resemble the Anaheim North scenario in which the tourism industry takes over, or will it be more like the Walleye Commons in which deregulation leads to economic collapse followed by regeneration. *(From Carpenter et al. 2002.)*

Northwoods Quilt

In the third scenario, recent retirees who relocated to the NHLD play an integral role in preserving the natural beauty that originally attracted them to the area. The lake associations to which they belong become effective fora for discussing management strategies. One practice adopted is to designate certain lakes for certain uses, such as power boating or canoeing. There's a patchwork of different kinds of ecosystems.

The retiree population expands and becomes more influential in the politics and economics of the NHLD. The economy diversifies because some retired professionals work part-time via travel or telecommuting. Resource conflict resolves in a multi-tiered system of regulations and incentives that allocates considerable power to lake associations.

By 2027, the NHLD is a mosaic of diverse ecosystem uses. The NHLD is buffered from ecological disturbance by the diversifying composition of its landscape.

Refugee Revolution

The final scenario presents an extreme situation. A plane flying over Chicago drops two tanks of radioactive dust, causing people to flee from the urban terrorism to the NHLD area. As a result, the population doubles in size and new businesses emerge. The government also turns to the region as a national resource for water, fish, deer, and even trees.

Terrorism makes urban life chaotic and dangerous. Many people abandon cities for rural areas. Owners of recreational properties in the NHLD move there to stay. Initially the infrastructure is severely stressed, but strong interventions by state and federal governments eventually create a viable economic base for a much more populous NHLD.

By 2027, working ecosystems producing water, cranberries, fish and game for market, and forest products dominate the NHLD landscape.

Engaging with Uncertain Futures

Particular events may seem implausible, but they're a class of event— a plausible category of what could happen. The final scenario, for example, is not predicting that Chicago will become a radioactive wasteland. It is only suggesting this as a possible future storyline in which the NHLD undergoes a sudden population increase due to some major, external event.

For this reason the scenarios should be considered together, not separately. They should be thought of as a set that provides us with a range
of insights on what makes a region vulnerable and what confers
resilience. Together, they present different dimensions of how things
might change (Carpenter, in press).

So what do these four scenarios suggest about the NHLD?

First, the region is vulnerable because of the low diversity of economic
opportunity and its openness to economic and political forces from outside the NHLD. Ironically, the traditional self-reliance that is valued by
many local residents may undermine the networking and collaboration
that could make the region less vulnerable to external influences.

Resilience is conferred by several features of the NHLD. One source
is the tribes, who intend to stay in the region come what may. Another
source of resilience is the capacity for renewal of the ecosystems of the
NHLD. Ecological breakdowns can occur due to poor stewardship of
shoreline habitats, biotic invasions, overharvesting, and so forth. However, the diversity of lakes and the capacity to manage different lakes
for different purposes provides a range of alternatives from which future
success might arise.

Key sources of innovation in the scenarios are the tribes and the
newly retired or semi-retired professionals who immigrate to the
region. The tribes are an important source of young people who want
to stay in the region. In addition, they diversify the perspectives on
resource management and the kinds of tourism opportunities in the
region. Incoming older residents bring different viewpoints on
resource management, different economic activities, and new prob-
lem-solving skills.

Unforeseen events might open up or close down different futures,
but it's the underlying mix of vulnerability, resilience and innovation
that will craft the region's future.

Can Scenarios Change the Future?

Will these scenarios change the future? In a sense, we will never know,
because there is only one NHLD and the scenarios are now in play
with no control or reference system to help us interpret the outcome
(Carpenter, in press).

However, we do know that these scenarios have stimulated debate and new thinking. In surveys undertaken after the release of the scenarios, most respondents hope that the future brings something like Northwoods Quilt or Walleye Commons. Against that hope, most respondents actually believed the future would most likely resemble Anaheim North given the existing trends at the time.

About 70 percent of respondents said they would like to become involved in a group working for desirable change in the NHLD. Although they are willing to act, most respondents believe that they have little influence on the future of the NHLD. Twenty-five percent said they will move away if the NHLD begins to change in undesirable ways.

Better networking is one key to building the adaptive capacity of the NHLD. The workshops that led to the writing of the scenarios have already formed new networks of contacts in the NHLD. More connections among key people and groups are necessary for adaptive change in the region (a theme that is explored further in the final case study on the Kristianstad Water Realm).

Substantial benefits could emerge from more frequent exchange of ideas between the innovative institutions in the region, such as the tribes, lake associations, and research organizations. A few interesting experiments in governance, collaboration, and ecosystem management are already underway in the NHLD. It will be important to share the results of these explorations.

The present already contains elements of all four scenarios, and the same is likely to be true of the future. Also, the future will contain many surprises that are not in the scenarios. Which scenario elements and what unforeseeable surprises will dominate the future? What parts of the past will people choose to carry in the future, and what parts of the past will be abandoned? What boundaries will be accepted by the people of the NHLD? What boundaries will people revolt against, and overcome? As the NHLD reorganizes, what new boundaries will be created? These questions will be answered over time, as the people of the NHLD act on the expectations and visions for the future.

Adaptability and transformability depend on the capacity of people to maintain or change the social-ecological system in which they live. Adaptability to upcoming challenges depends on human choices being made now. Better choices are likely if evolving changes are faced clearly and collaboratively, with minds open to the surprises to come.

Resilience and the NHLD

The residents of the NHLD are engaging with their future. They are imagining what the future may hold given changes in key uncertainties regarding population and ecological vulnerability. The very process by which they are doing this, through the creation of scenarios, is enhancing the region's adaptability and resilience by establishing networks, and encouraging the various actors in the system to explore the region's vulnerabilities, resilience, and sources of innovation.

It's interesting to note that a quarter of the NHLD survey respondents indicated that they would move out of the region if its natural values continue to decline. This begs the question: Move to where? To some other region rich in natural values that hasn't filled up yet?

5

Making Sense of Resilience:

How Do You Apply Resilience Thinking?

n the early 1980s, southeastern Australia was gripped by a major drought, the worst on record according to some. Farmers everywhere were facing bankruptcy as their paddocks turned to barren dust bowls. There was little grass, and the drought winds were blowing away what vestige of organic matter and topsoil was left.

On one farm near Canberra, the nation's capital, a third-generation sheep farmer named John Weatherstone came to a stark conclusion that despite applying what was considered best practice in farming, he was working the land too hard and it had lost its capacity to generate the produce that his business depended upon. He had two concerns; after the drought, would his farm start to recover (as opposed to staying degraded, or even getting worse)? And how long would it take?

The drought had exacerbated a host of problems such as poor soil, the death of trees, declines in biodiversity, and a loss of productivity. John concluded that if his farm, Lyndfield Park, was to have any future he needed to change what he was doing.

In the years following the drought John did his best to reduce the pressure his traditional farming practices were placing on the land: he reduced grazing pressure, minimized soil cultivation, cut back on farm chemicals, and planted more than eighty thousand trees and shrubs. Indeed, selling native trees, shrubs, and seeds has now become one of the main enterprises of Lyndfield Park. He stopped grazing sheep altogether and turned to cattle as he believed sheep grazing was exacting too great a toll on the soil. Much of what he was doing he needed to figure out for himself as he did it, because conservation farming hadn't been developed back then.

Farm forestry for timber products is also being developed at Lyndfield Park using a range of native trees, and John is running a number of trials of native species to test their growth and marketability. He's finding that the strength of his farm enterprise lies in the diversity of his projects so he's always willing to try new ideas. He's also found that diversity is the key to successful revegetation for wildlife. Without intending it, his farm is now a local hotspot for native animals, having more species of native birds on it than any other property in the area.

Over the last two decades the property has undergone a dramatic transformation as John has turned from grazing sheep to growing trees and grazing cattle. In the process he's turned around many of the threats undermining the viability of the farm, significantly increased its resilience to the stresses of drought, improved its financial performance, dramatically increased its capital value, and created an attractive and pleasant place in which to work.

Twenty years on and southeastern Australia is again gripped by drought, possibly worse than the episode of the early 1980s. Farmers are again going broke. This time, however, John Weatherstone's farm is coping much better than his neighbors.

He writes: "we're praying for rain, yet, as I look out over Lyndfield Park, I have reason to hope. Most of the property is protected by trees, shrubs and perennial grasses. All of the vegetation is dry, some of it is stressed and some of it will probably die. However, most of it is still intact and it's protecting that valuable mantle of top soil that is the life blood of our many enterprises. On many properties that precious mantle is being lost to the wind. I'm confident that when the rains come, whenever that may be, the resilience and health of our farm will enable it to recover" (Weatherstone 2003).

We begin this chapter with this example because it underscores several of the themes that have been covered in previous chapters and case studies, and sets out one man's efforts to manage for resilience. Before we discuss how this is relevant, let us briefly reiterate where we've come from.

The Building Blocks of Resilience Thinking

One of the central ideas of resilience thinking is that social-ecological systems have multiple regimes (stable states) that are separated by thresholds. The metaphor connected with this model is a ball in a basin (chapter 3). The ball is the current state of the social-ecological system, and the basin is the set of possible states the system can be in and still have the same structure and function. Beyond some limit (the edge of the basin), the feedbacks that drive the system change and the system tends toward a different equilibrium. The system in this new basin has a different structure and function. The system (the ball) is said to have crossed a threshold into a new basin of attraction. In this metaphor, resilience is all about the distance between the ball and the edge (threshold) of the basin, and the size and shape of the basin of attraction.

The other building block of resilience thinking is the notion of adaptive cycles (chapter 4). This sets out how social-ecological systems behave over time going through cycles of growth and conservation (the fore loop) followed by release and renewal (the back loop). The fore loop is characterized by the slow accumulation of capital and potential, by stability and conservation. The back loop is characterized by uncertainty, novelty, and experimentation. Different cycles operate at different scales and the linkages across scales are very important. What happens at one scale can influence or even drive what's happening at other scales.

Putting these pieces together gives us a basis for resilience thinking, an approach to appreciating what's driving and configuring the enterprise or organization that is of most interest to us. It's all about seeing a farm/family/business/region as a complex adaptive system that's constantly changing and adapting to a changing world.

Resilience is the capacity of this system to absorb change and disturbances, and still retain its basic structure and function—its identity. Resilience thinking is about envisaging a system in relation to thresholds. Is it approaching a threshold beyond which it will be in a new regime? What forces—economic, social, and environmental—are driving the system toward this threshold?

Using the ball in a basin metaphor, can you, through your management, influence either the shape of the basin, or the position of the system within the basin?

And now try to envisage the system as a set of linked adaptive cycles. Where does the system (at the scale that you're most interested in) lie in the adaptive cycle? What is happening at the scales above and below the scale of your interest? What are the linkages with these scales? What's driving your system and where is it headed?

Resilience thinking is a way of looking at the world. It's about seeing systems, linkages, thresholds, and cycles in the things that are important to us and in the things that drive them. It's about understanding and embracing change, as opposed to striving for constancy. It arose from attempts to understand how and why ecological systems change over time but it has evolved to encompass a lot more.

Anyone Can Do It

Taken as one big pill, resilience thinking can be a bit difficult to swallow. It contains several interlinked concepts and involves a new way of looking at things. However, when applied in context to the system you are interested in, it provides valuable insight. What's more, it doesn't need a Ph.D. to be applied. Indeed, that's why we began this chapter with the story of Lyndfield Park.

John Weatherstone is a farmer, not a trained ecologist. He hasn't undertaken a formal resilience analysis of his farm system. Resilience thinking wasn't even a developed body of theory when he transformed his farm. His change of direction came as a gut reaction in a time of crisis. His understanding of resilience wasn't couched in terms of thresholds and adaptive cycles, and yet his assessment is completely in keeping with them. His definition of resilience is that of the land's capacity to recover from the impact of prolonged drought, and to continue to provide the ecosystem goods and services that underpin his enterprises. That's how we define it as well.

This is a good example of homegrown resilience thinking, and it goes to show that anyone can do it. You don't need a detailed appreciation of thresholds and adaptive cycles to apply it. You *do* need to see your enterprise as part of a broader interlinked system, be able to identify the important processes and variables that underpin your operation, and have the capacity to ask the appropriate questions. And you need the capacity to implement change.

Having said that, thresholds and adaptive cycles at different linked scales allow for added insights on what's happened at Lyndfield Park and how it might relate to other farms, other catchments, and other scales.

John Weatherstone's epiphany in the drought of the early 1980s was that applying traditional farming practice was driving his farm system over a threshold. To continue in that direction was to destroy the enterprise he and the generations before had worked so hard to set up. Prior to the drought, Lyndfield Park was a model of (traditional) farm efficiency applying the latest in technology and economic modeling. A couple of decades earlier an accelerated program of pasture improvement saw significant increases in productivity with stock carrying capacity more than doubling in the 1970s. There was also quite a lot of cash cropping being done. It was very much an efficiency-driven approach based on growth and optimization.

He didn't realize it at the time but that growth was at the expense of parts of the natural systems that were renewing and revitalizing his soil, and providing cover in times of drought. However, in good times when things are going well it is not something you stop and question. The farm had moved into a conservation (K) phase of the adaptive cycle with resources increasingly committed by the success of the gains in efficiency. John was investing in more intensive agriculture using higher chemical inputs and higher stocking rates. He was working the land harder to generate more income to cover the higher level of investment.

Then a massive shock hit the system. Not just the ecosystem, the social-ecological system that was the agro-ecosystem plus John and his family. And this social-ecological system went into the release phase of the cycle.

John described it thus: "In a time of crisis, it's not uncommon for a farmer to become introspective. We might not say much, but when the land you've been working for so long suddenly dries up and blows away, taking with it your financial security, it makes you think about things" (Weatherstone 2003).

"Thinking about things" and completely rearranging your business is something that doesn't happen easily. The release phase in this particular adaptive cycle is a good example of creative destruction in which traditional ways and linkages are overturned in favor of new ideas and different approaches. The crisis proved to be the opening of a window of opportunity for John and his family. In the reorganization phase he had some sheep and he introduced four new ways of farming: cattle, tree seeds (that he collected on and off the farm), shrub and tree seedlings, and tree plantations for timber. Within a few years he got rid of all his sheep (a business that had been in the family for generations), gave up his seedling nursery business and concentrated on cattle, tree seeds, and timber trees.

Scales and Cycles

John's story can also be viewed in terms of farming activity at larger scales. John is more productive, better insulated from droughts, and a happier farmer these days. However, he's not representative of an average farmer, and is regarded as a bit of a maverick. Other farmers in the region are still sticking to traditional models of farming and attempting to optimize production. In so doing they continue to increase their exposure to episodes of drought and other climatic extremes. Drought aid in the region is once again the order of the day, as many farmers cross the threshold to bankruptcy.

This is the status quo, and it is not easily changed. Traditional farming based on increasing efficiency is still advocated and encouraged at state and federal levels of government. Farmers are supported and educated on pasture improvement, land intensification, and controlling variability with chemicals and machines. John's approach is tolerated, even celebrated in some circles, but not yet actively fostered by the broader industry or by government.

All farms are influenced by what is happening at the larger scales of the industry, the state and the country. Though individual farms might be hit by a drought and enter a back loop, the average farm in the region emerges in the next cycle operating in the same way; sometimes with a new owner following the economic demise and departure of the previous one. The state of the system at a larger scale (the farming region) affects the trajectory of a smaller scale cycle, in this case encouraging it to repeat itself (a systems process known as "memory"). Sometimes this is in the form of drought subsidies from government that allow some farmers to emerge from a drought and start again—in the same way.

However, linkages across scales can work both ways. For example, a few farmers are taking steps in a similar direction that John has taken. There are probably a few more asking questions as they batten down during the current drought. If enough farmers in the region were to adopt a more resilient farming model across the catchment then their activities might reverberate up to the scale of the region and result in a change in state or federal regulations that foster such farming practices in other regions. When changes on a smaller scale cycle combine they can initiate a revolt at scales above, precipitating change and reorganization at these higher scales.

BOX 8 Seeing the World through a Resilience Lens

Explaining resilience thinking to someone who understands resource management and sustainability is an interesting experience because they often get the various bits that make up a resilience framework without immediately appreciating their consequences.

However, if they carry the messages of resilience with them, they start to look at the world in a new way. The penny drops as they begin to interpret events around them, past and present. Thresholds and adaptive cycles are all around us operating over many scales in time and space. If you are aware of their existence and their implications for how social-ecological systems function and change over time, you start looking for them. And as you identify them you begin to appreciate why a system is behaving as it does, and what lies behind certain outcomes.

For example, why did civilization crumble on Easter Island? Why, when vast resources are being thrown at it, are we unable to halt the biological decline of the Everglades? How did we become so dependent on pesticide and herbicide use? Why are we facing a worldwide threat from antibiotic-resistant bacteria? What brought about the end of the Soviet Union? Why did the Roman Empire collapse?

Various theories have been proposed in answer to each of these questions. However, if you think through the situations behind each of them, it's apparent that whatever the form of business as usual, it didn't work. There was a failure in each situation to appreciate the dynamics of a complex adaptive system. Thresholds, resilience (or a lack of resilience), and adaptive cycles can be identified as playing a critical role in each of these events.

Operationalizing resilience thinking is, in part, about getting people to cross a mental threshold into a systems mind space in which systems with multiple stable states and adaptive cycles make sense. Cross this particular threshold of understanding and the world takes on a different light.

Operationalizing a Resilience Approach

The story of John Weatherstone's farm is one example of how resilience thinking might manifest itself. How might managers of the broader catchment apply a resilience framework? What about leaders and policy makers at state, national, and international levels? Or, what about the owner of a business or a policy maker for an industry? How might resilience thinking apply to different situations and different scales in time and space? The five case studies presented in this book provide some ideas on the relevance of resilience thinking to different regions around the world.

In general terms it is all about appreciating the social-ecological system that you are interested in (and are a part of) as a complex adaptive system, and defining its key attributes. What are the key slow variables that drive this system? As these variables changing, are there thresholds beyond which the system will behave in different ways? If so, where are these thresholds? Thresholds are defined by changes in feedbacks, so which important feedbacks in the system are likely to change under certain conditions? What phase of the adaptive cycle is the system moving through? What is happening in the adaptive cycles above and below the particular scale you are interested in? What are the linkages between scales?

These are all big questions and they don't come with simple answers. Whole books have been written on how these questions might be approached, and there are no definitive solutions. It is beyond the scope of this book to provide a detailed approach to how resilience thinking might be operationalized. The development of a *Practitioners Workbook* on how to undertake a resilience analysis is a work-in-progress project of the Resilience Alliance, due for completion in late 2006. In the interim, there is a body of ideas, case studies, and guides contained in our reading list.

However, the very effort to frame (and then attempt to answer) these questions is a major step toward defining resilience and sustainability. You start to appreciate that there are certain variables changing in the system you are a part of that you can't afford to ignore. You begin to look for thresholds and start asking questions about the trajectory the system is on.

The question of sustainability becomes one of what kind of system regime do you want the system to be in? (Or, perhaps, what kinds of regimes do you want to avoid being in?) The concept of thresholds is critical to this question. If the system is on a trajectory that will carry it over a threshold where it operates differently (and in a way you don't want), then the question of sustainability is really how do you manage the system to avoid crossing the threshold? Or if the system is in a state where it is not giving you the goods and services you need, the question becomes one of how do you move the system over a threshold into a different regime (basin of attraction)? If you are unsure about system regimes and possible thresholds, then how can you build resilience and adaptive capacity, generally, to increase the ability the cope with external shocks?

An understanding of the history of the system is a good starting point in attempting to recognize adaptive cycles and system regimes. Consider the histories of our five case studies. In each there are fore loops of growth and predictability, and back loops of creative destruction and new beginnings. An appreciation of where and how adaptive cycles are operating in a system leads you to look at linkages across scales and how they impact on the scale of your interest. It also encourages you to choose when you might attempt to influence a system by seizing windows of opportunity when they open.

Adaptability and Transformability

Having analyzed the social-ecological system in which you are interested, appreciated the key slow variables that are driving it, recognized real or suspected thresholds, and identified cycles and cross-scale connections, you now need to consider how you might work with your system's resilience.

The capacity of the actors in a system to manage the system's resilience is known as adaptability (also referred to as adaptive capacity). This might be by moving thresholds, moving the current state of the system away from or toward a threshold, or making a threshold more difficult or easy to reach.

In the Everglades adaptability might be defined as the capacity of the managers of the system to regulate phosphorus levels in the water, or the impacts of those levels on the ecosystem. In the Goulburn-Broken Catchment adaptability might relate to the capacity of catchment managers to facilitate extensive revegetation of the landscape to reduce water from reaching the groundwater. In the Caribbean it relates to the capacity to reinstate key functional groups of organisms—to get some key species back into the ecosystem.

It might be that it's impossible to manage the threshold or the system's trajectory if it is stuck in an undesirable basin of attraction. In such cases it might be more appropriate to consider transforming the very nature of the system.

The Cost of Resilience

Loss of resilience happens, most often, unwittingly. However, even when its loss becomes apparent (when thresholds are recognized) it is still usually ignored or downplayed. Why is that? The main reason is that maintaining resilience comes at a cost.

Strategies to increase response diversity (introduce "redundancies"), reduce efficiencies, and stay away from maximum yield states all carry costs, usually in the form of lost short-term opportunity gains. As long as things don't go wrong, it's more profitable to run down resilience. The longer this goes on, the more resilience is reduced, the higher the likelihood of an unwanted regime shift.

Managing resilience, therefore, comes down to assessing the short-term profit losses associated with maintaining or enhancing resilience, against the long-term benefits of not undergoing a regime shift.

short-term cost of foregone extra yield

vs.

cost of being in an alternate regime

(cost x probability of regime shift)

This is a difficult balance to assess because there are few systems in which we have the detailed knowledge to be able to assign the costs and benefits.

As more people adopt a resilience approach, and as the body of information grows—especially about the consequences of regime shifts—the better we'll be able to make such assessments. For now, it is important to recognize that there are short-term costs, and that they need to be considered in any operational program.

General and Specified Resilience

When considering how you might manage a system's resilience there is usually an emphasis on appreciating specific threats to a system. The approach is to define the system in terms of thresholds, and this means attempting to understand the key slow variables that are configuring it—the ones that might exhibit threshold effects. This might be the level of the groundwater (e.g., the Goulburn-Broken Catchment) or the level of phosphorus in the sediment (e.g., the Everglades). Once you've identified these key variables, the question of resilience becomes specified—for example, "the resilience of the Everglades vegetation to fires and droughts (as phosphate levels increase)".

Not surprisingly, this is referred to as specified or targeted resilience, the resilience "of what, to what." If you can perform this type of analysis on the system then you have already come a long way from a business as usual approach. However, by itself it is not enough. Resilience thinking needs to go beyond managing for specific variables and specific disturbances.

A common theme in this book is that optimizing anything, including specified resilience, comes at the cost of limiting your capacity to respond to unforeseen shocks and disturbances. Optimizing for one form of resilience can reduce other forms of resilience.

Managing for specified resilience is important, but so too is maintaining the general capacities of a social-ecological system that allow it to absorb unforeseen disturbances—that is, general resilience.

What confers general resilience? Studies of a variety of social-ecological systems, including the five presented here, suggest three factors that probably play an important role in maintaining it are diversity, modularity, and the tightness of feedbacks—all key features identified for ecosystems by Levin (1999).

Diversity refers to variety in the number of species, people, and institutions that exist in a social-ecological system. It includes both functional and response diversity (recall the discussion of the Caribbean coral reefs). The more variations available to respond to a shock, the greater the ability to absorb the shock. Diversity relates to flexibility and keeping your options open. A lack of diversity limits options and reduces your capacity to respond to disturbances. Increasing efficiency (optimization) inevitably leads to a reduction in diversity.

Modularity relates to the manner in which the components that make up a system are linked. Highly connected systems (lots of links between all components) means shocks tend to travel rapidly through the whole system. Systems with subgroups of components that are strongly linked internally, but only loosely connected to each other, have a modular structure. A degree of modularity in the system allows individual modules to keep functioning when loosely linked modules fail, and the system as a whole has a chance to self-organize and therefore a greater capacity to absorb shocks.

Tightness of feedbacks refers to how quickly and strongly the consequences of a change in one part of the system are felt and responded to in other parts. Institutions and social networks play key roles in determining tightness of feedbacks. Centralized governance and globalization can weaken feedbacks. As feedbacks lengthen, there is an increased chance of crossing a threshold without detecting it in a timely fashion.

In the case of the Goulburn-Broken Catchment, the response to the underground flood of the 1970s was the establishment of a new institution of a catchment management authority that devolved responsibility for responding to an environmental threat from central government to a regionally based institution. The catchment authority was strongly linked to local social networks, and played an important role in responding to the threat of rising groundwater. The creation of the authority increased the region's resilience because it increased the tightness of feedbacks.

In the Northern Highlands Lake District, the development of scenarios served to enhance the resilience of that region in two ways. It served to build and strengthen social networks that allowed for a more rapid response to changes, and the scenarios provided indicators, advanced warnings, of unwanted changes, thereby shortening the feedback times of such changes to the community.

In our next case study on the Kristianstads Vattenrike (or Kristianstads Water Realm), the central element of resilience in that region grew out of an evolving social network that came together in response to a growing sense of environmental crisis.

Some Implications for Policy and Management

Rather than focusing on the need to control natural variability and to maintain the system in some perceived optimal state, a resilience approach to management and governance would instead focus on alternate system regimes and thresholds and the capacity to avoid or manage them. How can this general statement be made operational in terms of policy and management?

The policy implications of a resilience framework will be explored in some detail in the *Practitioners Workbook* currently under preparation by the Resilience Alliance. For now, however, here are some considerations for policy that should be borne in mind. In the final chapter we will refocus on these considerations in the context of what a resilient world might be like.

- You can't manage ecosystems or social systems in isolation. Their strong interactions mean that feedbacks between them must be taken into account.
- When taking account of resilience it's important to know what

phase of the adaptive cycle a system is in. Is it nearing a change to a different phase? What kinds of interventions are appropriate, or inappropriate, in the current phase?

- An understanding of what is happening in the scales above and below the scale at which you are working is important. What effect do these scales exert over the scale in which you are interested?
- It's also important to identify the key (slow) controlling variables that may (or do) have threshold effects. Look for, and understand the drivers of, slowly changing variables (in the ecosystem and in the social system).
- Identify any possible alternate regimes for the system, based on the controlling (slow) variables. Crossing in to an alternate regime will usually mean that the supply of goods and services from the system will alter.
- Be aware that simplifying the system for increased efficiency reduces the system's diversity of possible responses to disturbance, and the system becomes more vulnerable to stresses and shocks.
- Identify the key points for intervention that can avoid undesirable alternate regimes. This amounts to either changing the positions of thresholds (by identifying and managing the system attributes that determine them) or changing the trajectory of the system.
- When help for communities/industries in trouble is warranted, devise subsidies *for* change, rather than subsidies *not to* change.
- Invest in building adaptability (social capacity—trust, leadership, networks) and promote (do not hinder) experimentation and learning.
- Design or modify existing governance structures so that key intervention points can be addressed at the appropriate scales and times.
- Acknowledge that there is a cost to maintaining resilience. It comes down to a trade-off between foregone extra profits in the short term, and long-term persistence and reduced costs from crisis management.
- When the system has already moved into an undesirable regime (where the endpoint, or equilibrium, is unacceptable, and efforts

are being made to keep away from it) there may come a time when adaptation is no longer socially or economically feasible. When transformation is the only option, the sooner it is recognized, accepted and acted on, the lower are the transaction costs and the higher the likelihood of success.

Key Points on Resilience Thinking

- You don't need a degree in science to apply resilience thinking; you do need a capacity to look at a social-ecological system as a whole and from different perspectives and at different scales.
- When managing for resilience you need to consider two types of resilience: resilience to disturbances that you are aware of (specified resilience), and resilience to disturbances that you haven't even thought of (general resilience).
- Adaptability describes the capacity of actors in a social-ecological system to influence the system's trajectory (relative to a threshold) and the positions of thresholds.

Building Resilience in the Wetlands:
The Kristianstads Vattenrike, Sweden

I n 1975, the people of Kristianstad had cause to believe the health of their local environment was about to improve with international recognition of their local wetland. Ten years later, however, and their hopes were turning to despair as the local environment continued to fall apart. Recognition by the outside world didn't make much of a difference. If their local environment was going to be managed sustainably, it looked like they would have to carry the responsibility themselves. The need was there and the region was primed for change.

The international recognition that Kristianstad had pinned its hopes on in 1975 related to the Convention on Wetlands of International Importance. A thirty-five-kilometer stretch of wetlands along the nearby Helgeå River had just been declared a Ramsar Convention Site. This meant that the wetland would be protected from further exploitation and managed in accord with a comprehensive conservation plan explicitly formulated to look after the natural values of Ramsar sites.

It was hoped the announcement would herald a reversal in the trend of environmental decline that had been evident all around the city for many years. Unfortunately, ten years later it seemed the protection afforded by the declaration was illusory. Surveys and observations by several local groups in and around the Ramsar site revealed the area's natural values were still in decline. Bird populations were decreasing; the lake was experiencing

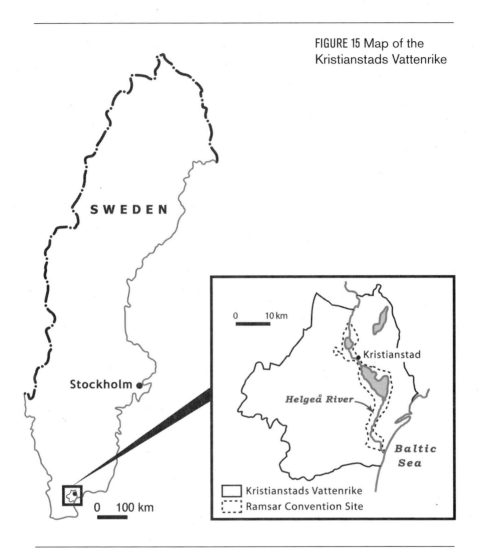

FIGURE 15 Map of the Kristianstads Vattenrike

increasing eutrophication and was being clogged by plant growth. There was growing concern that if international certification couldn't preserve the cultural and natural values of the area then perhaps nothing could.

Sometimes the perception of a crisis can catalyze change every bit as effectively as a crisis actually happening, and the climate in Kristianstad in the 1980s was ripe for a revolutionary new approach toward looking after the region's natural and cultural resources.

IMAGE 15

Kristianstads Vatten-rike with the Helgeå River and the Municipality of Kristianstad, in the foreground, and the Hanö Bay of the Baltic Sea in the background. (*Courtesy of Patrik Olofsson [From Olsson et al., 2004.]*)

The Kristianstads Vattenrike in a Nutshell

Kristianstads Vattenrike (KV) roughly translates as the "Kristianstad Water Realm," however *rike* also means "riches." This double meaning is quite appropriate as the name both defines the catchment area and reflects its rich natural values.

The KV is a semi-urban area of high biological and cultural importance in southeastern Sweden. It lies within the Municipality of Kristianstad and covers around 1,100 square kilometers of the Helgeå River catchment. Its southeastern boundary is defined by Hanö Bay.

KV includes Sweden's largest areas of flooded meadows used for grazing and haymaking. Many of the unique values of the area depend on these activities and the annual flooding of the Helgeå River.

At the heart of the KV lies Kristianstad, a city of some 28,600 inhabitants. The broader municipality supports a population of 75,000 people. Much of the KV is agricultural land with the sandy and clay soils around Kristianstad being important for agricultural production. The region is one of the most productive in Sweden.

Important natural habitats in the KV apart from the flooded meadows include large beech forests, wet forests, willow bushes, and sandy grasslands. Each habitat supports a range of unique flora and fauna. The area also holds the largest groundwater reserve in northern Europe.

The KV provides a range of important ecosystem services including the filtering of nutrients from water flowing to the coastal area of the Baltic Sea, recreational spaces, provision of significant habitat for a range

IMAGE 16

The Vramsån River, a tributary to the Helgeå River.
(*Courtesy of Carl Folke [From Olsson et al., 2004.]*)

of wildlife including the symbolic white stork, and maintaining the cultural and agricultural heritage of the landscape.

The area was designated to have international importance by the Ramsar Convention on Wetlands in 1975, and was recently accepted by UNESCO to become a Man and Biosphere Reserve.

A Social-Ecological History of the KV

People have lived in the northeastern part of Skåne (which we now know as the municipality of Kristianstad) for thousands of years. Many prehistoric finds testify to good settlement conditions here—forests with game to hunt, water for fishing and transport, and fertile soil to be cultivated. All these features, and the battles to win them, have left their mark on where the town now stands.

During the Middle Ages, when Skåne was still part of Denmark, the local towns of Åhus and Vä were important centers for trade and religion. These were uneasy times with Denmark and Sweden constantly involved in bitter battles. In the early 1600s, northeastern Skåne was repeatedly invaded by Swedish infantry who plundered, pillaged, and burnt everything in their path.

The Danish King, Christian IV, decided he needed a fortress to better secure his eastern border with Sweden. In 1614, he had a fortress built at Allön, in the heart of the almost impenetrable wetlands near the River Helgeå. During a visit the following year, the king decided that the town should be named after him, and was called Christianstad. However, as these things often go, the region came under Swedish rule some forty years later. Over the following 350 years, the town of Kristianstad developed into a center for trade, culture, education, and administration. And over this period, the landscape has been developed and tamed to increase the amount of arable land and control flooding.

Controlling the Water

As with so many regions around the world (such as the Everglades), a major part of this development involved the control of water levels and flows. One particular event that changed the area was the digging of a drainage ditch to the sea by local farmers in 1774 to prevent the annual flood from damaging their land. Unfortunately the spring flood of 1775 was so severe that it transformed the ditch into a new channel from the Helgeå River to the sea. This venture lowered the water level in the water system by more than half a meter.

The lakes, being shallow in the first place, were particularly sensitive to this lowering of the water level and have become more vulnerable to eutrophication. Both Lake Araslövssjön and Lake Hammarsjön are under constant threat of becoming overgrown by reeds and other plants.

The wetlands came under further threat with the building of embankments and dredging projects to control the river in the 1940s. The aim of this work was to speed the flow of water through the wetlands and prevent flooding, but the impact was to lower the water system by an additional third of a meter. Several small lakes between Lake Hammarsjön and the sea disappeared in the process. Growing urban sprawl and the construction of roads further fragmented the natural landscape.

The development of the river and the wetland was having a significant impact on the natural values of the region. It was also coming with an increasing social cost. Untreated sewage from industry and households was being poured into the river and water quality was declining. Public complaints were steadily rising during the first half of the twentieth Century and in 1941 the City of Kristianstad was forced to stop taking its drinking water from the Helgeå River.

Fertilizer use following World War II allowed for the intensification of agriculture at the periphery of the wetlands of the lower Helgeå. This not only increased the threat of eutrophication of the lakes but also threatened the groundwater with contamination by nitrates and biocides.

In the summer of 1964 there was a massive fish kill in the lower fifty kilometers of the river. Testing showed high levels of organic pollution and low levels of oxygen in the water. It is believed that this event wiped out the river's population of the rare European catfish.

The resentment of the broader community at how their environment was being treated began bubbling to the surface around about this time and open conflicts began to occur. If anything was going to inflame the situation it was the decision in 1966 by the local municipality to establish a garbage dump on the wet grasslands nearest to Kristianstad. Despite protests from local and national conservation interests that emphasized the biological and aesthetic values of the meadows, they were transformed into a city dump.

On the heels of this decision came plans to embank wet grasslands farther south of the city to open up permanent agricultural land. This prompted renewed protests among conservation interests who produced an inventory of natural values of the region. In response, the county administrative board decided to protect 150 hectares of these cultivated wet grasslands by establishing a nature reserve.

In 1974, the Municipality of Kristianstad took on the project of restoring two local lakes by removing the reeds and other plants that were choking them. In the following year, a thirty-five-kilometer stretch of the wetlands along the lower Helgeå River was declared a Ramsar Convention Site. Many thought this external intervention would be a turning point, but it was not the case.

Sustaining the Flooded Meadows

Despite several conservation efforts, studies undertaken during the 1980s indicated that the values of the lower Helgeå River and the

Ramsar wetland continued to decline. This was linked to the fact that flooded meadows used for haymaking and grazing had decreased dramatically.

Haymaking and grazing of flooded meadows had almost ceased in Sweden around the end of the nineteenth century, being replaced by the cultivation of fodder on arable land. However, a considerable part of the wetlands of the lower Helgeå River, some 1200 hectares, survived these reformations and rationalizations of agriculture.

In the 1980s, these remaining wetlands became threatened due to the abandonment of haymaking and grazing. If mowing ceases, the flooded meadows are overgrown by reed, sedge, and willow, as a stage in the ecological succession from flooded meadow to forest. To maintain the wetland, therefore, the grazing and mowing of the flooded meadows needed to continue. These flooded meadows are not an ecologically stable state. The desired regime of this social-ecological system is a cultural landscape with cultivation an integral component. Without the grazing management the ecosystem progresses to its stable equilibrium, a wooded system without flooded meadows (and without the wildlife dependent on those flooded meadows).

During this time, a curator of natural history in the Kristianstads County Museum named Sven-Erik Magnusson was actively collecting information on the history and dynamics of this cultural landscape of which the wetlands were one embedded part. He acquired an understanding of how local agricultural practices had shaped the landscape and ecosystems of the lower Helgeå River over thousands of years. He also developed a number of ways to communicate this understanding to the public and various stakeholder groups. One way involved the establishment of an outdoor museum (Utemuseum) in which visitors could interpret the wetland landscape as they walked through it.

Magnusson began studying the Ramsar Convention Site and found that where grazing and haymaking were still practiced the unique cultural values (such as traditional forms of agriculture) and natural values (such as waterfowl populations) were being maintained; however, where they had been abandoned, these values were in decline. In the mid-1980s he began working with members of the Bird Society of Northeastern Scania (BSNES) who were able to provide bird inventories for the wider region over longer time scales (going back to the 1950s).

By pooling their experience and knowledge, Magnusson and the

BSNES were able to convince the County Administrative Board and National Forestry Board, the bodies responsible for managing the flooded meadows in protected areas, that these ecosystems needed more than just protection, they required active management and a reinstitution of the traditional agriculture to sustain the natural values of these areas. In so doing, the cultural history and continuous use of the wetlands for grazing and haymaking was linked to the ecological qualities for maintaining a rich bird habitat.

In short, they needed high adaptability in order to manage the resilience of the flooded meadow regime and to reduce the strong tendency of the ecosystem to move to its equilibrium state of a forest.

Establishing the Ecomuseum Kristianstads Vattenrike

It was through experiences such as these that Magnusson learned the importance of linking the knowledge and experience of actors at different organizational levels. He was aware of different people and groups operating at different levels in a number of activities. This included creating inventories (natural and cultural), running monitoring programs, carrying out restoration, attempting to develop improved land use and management practices. He was also aware that the groups undertaking these activities were frequently unaware of each other. He realized that there was a need to gather these activities into one concept, and that concept was the Ecomuseum Kristianstads Vattenrike (EKV). The EKV was basically a forum in which the different actors and groups with a stake in Kristianstads Vattenrike were given an opportunity to meet and exchange ideas and values, and develop a shared understanding and vision for the future of the region.

To garner support for the EKV, Magnusson focused on specific individuals in key organizations that had some interest in Kristianstad and its surrounding wetlands. This included researchers from local universities, officials from the World Wildlife Fund (WWF), a former president of the Kristianstad Tourism Board and the director of Sweden's National Museum of Natural History. With their support and participation, the EKV took form as a body that would play an important role in conflict resolution, information sharing, and coordination.

As the concept grew it engaged support and involvement from a number of other groups including the County Administrative Board, the

BSNES, and environmental and farmer associations. The individuals representing these groups became the nodes of an emerging social network.

In 1988 the municipal executive board acknowledged the growing environmental threat to the wetlands and the potential for a body like the EKV to assist in managing the wetlands as a valuable resource for recreation, tourism, biodiversity and water purification. The following year the Municipality of Kristianstad assumed responsibility for the running of the EKV.

The initial funding for the EKV depended on the willingness of various actors to contribute to the process. Different parts of the EKV project appealed to different sponsors, and all sponsors made their support conditional on broader participation by other sponsors. The municipality, for example, provided funding for a person to start an EKV project on restoring the flooded meadows on the condition that the County Administrative Board funded an adviser. The WWF was happy to fund aspects of the fencing and clearing if the municipality provided the administration.

IMAGE 17

Dialogue was a key factor in the transformation of the governance of the wetland landscape. The photo shows three key individuals representing three different managerial levels: the County Administrative Board, the Municipality of Kristianstad (EKV), and the farmers. *(Courtesy of Carl Folke [From Olsson et al., 2004.])*

Making a Difference

The EKV has subsequently developed into a flexible and collaborative network with representatives from several levels of society, from local to international (Olsson et al. 2004). It has been involved in numerous interventions since its inception. The mapping of the flooded meadows in 1989 provided valuable information for how the wetland ecosystems might be prevented from developing to an undesirable state. The response to the threat of the flooded meadows becoming overgrown was to create social structures and processes to secure their continued cultivation. This was also important for enhancing the declining bird populations that depend on flooded meadows.

In collaboration with the WWF, bird societies, the Swedish EPA, and the County Administrative Board, the EKV compiled a number of ongoing inventories including the mapping of reserves, cultivated areas, bird populations, and nutrient levels. The results were communicated to a variety of actors, including the general public, using a wide range of methods. Creating such feedback loops is a prerequisite for managing complex systems sustainably.

The EKV maintains a close collaborative relationship with the farmers, making use of their knowledge and understanding of agricultural practices that have often been developed and passed on from generation to generation. An example is adjusting grazing on flooded meadows in relation to biodiversity. If cattle are the only grazers, the ground takes on a tussocky surface. If grazed by horses it develops a smooth and even surface. Mowing the wetlands can simulate this effect. Some bird species are dependent on a mix of the two types of surfaces. The EKV used a variety of mapping exercises of these habitat types to increase the farmers' knowledge of the unique values of their land in a larger context.

The EKV has made the wetland landscape area more accessible to the public and has established thirteen information sites throughout the wetlands. More than a hundred and fifty thousand people visit these sites each year.

Although the initial work of the EKV focused on the flooded meadows, it has expanded to include other aspects of the region. The association has been involved in the reintroduction of the white stork and European catfish, dealt with the problem of increasing numbers of geese and cranes, restored tributaries of the Helgeå River, and assisted in the

IMAGE 18

Cranes in the KV spend their days feeding in farmers' fields and serve as a drawcard for ecotourists. Rather than chasing the cranes away, the farmers agreed to sacrifice a portion of their production and receive compensation. *Photo credit: ©S-E. Magnusson, Kristianstads Vattenrike Biosphere Reserve.*

management of floods. And as these projects have progressed the network of collaborators in the EKV has grown.

The manner in which the EKV dealt with the problem of cranes serves as a good example of how it operates. Increasing numbers of cranes were damaging farmers' plantations. However the cranes enhanced the value for recreation and eco-tourism. A potential conflict was forestalled when the EKV talked with the affected farmers and established a crane group with representatives from different stakeholder groups. Rather than chasing away the cranes, the farmers agreed to sacrifice pieces of their land and receive compensation for losses in harvest.

The existence of collaborative networks across stakeholder groups at the municipal level was crucial for this process to have taken place. Mutual trust already existed and this new problem/crisis/opportunity was turned into a win-win situation.

Diversity Assists in Managing for Complexity

The success of the EKV in managing the wetland ecosystems of the lower Helgeå River has a lot to do with its structure and function. The EKV is part of the municipality's organization and reports to the municipality board. However, it is not an authority and has no power to make or enforce formal rules. By serving as a forum that brings together individuals and organizations to discuss emerging issues, build consensus, provide feedback, and share views, the EKV serves a valuable role in building trust and enhancing the resilience of the social-ecological system that is the KV. The very diversity of its membership is a significant part of its effectiveness in dealing with the complexity of the system.

Depending on the type of problem arising, various affected people are gathered by the EKV to be part of the process of solving the problem. It acts as a facilitator and coordinator in such an event. The actors are part of the planning, implementing, monitoring, and evaluating phases of the learning process. Management practices emerge and are revised as they are implemented.

In addition to this there are regular meetings of a reference group within the EKV to produce mechanisms for conflict management. The idea is to bring together representatives of any group involved in activities with links to the KV. This builds trust among the representatives, an essential component to the success of the collaboration process. If discussions on collaboration are only initiated once a conflict has arisen it's much more difficult to reach consensus.

Formal agreements and action programs emerge from these collaborative processes. These in turn lead to a change in behavior and practices in order to improve the management of the wetland ecosystems. The success of the EKV over time would suggest that this approach, often referred to as "open institutions," produces faster and more long-lasting results than making authorities develop rules that force people to change their behavior.

There are several lessons in the KV that contribute to our understanding of resilience in social-ecological systems. The first is that the imposition from the outside of a set of rules to protect an ecosystem, such as the establishment of the Ramsar Convention Site, will not ensure the natural qualities of a region will be preserved over time. One size never fits all, and an understanding of local history and culture needs to be integrated into the management if local values are to be looked after. For that to happen, local people need to be part of the process.

Another lesson is that the processes and values that influence the management of an area operate over many scales—local, regional, national, and international—so for an organization to meaningfully deal with this complexity it needs to include representatives from each of these levels in the social network. These representatives need to be engaged in such a way that they will both contribute to the governance of the system and share that responsibility with other representatives and feedback to their relevant organizations.

Finally, the formation of the EKV took place because several organizations with a stake in the KV were prepared to contribute to a shared vision and build consensus on how the KV might be managed. However, in its earliest stages, the formation of the EKV was catalyzed by an individual who brought these various actors together. Leadership is a crucial component in building adaptability and transformability.

Transforming the Social Domain

Like the Goulburn-Broken Catchment, the ecosystem of the KV social-ecological system is already in a basin of attraction with a stable equilibrium state that nobody wants. In the Goulburn-Broken case it seems that the efforts to keep the system from this state (groundwater and salt reaching the surface) can never be enough as long as the current (preferred) system of irrigated farming continues. Adaptability can't help; transformability is ultimately the only way out.

In the KV, however, the adaptive capacity that has been built in the transformed social part of the social-ecological system is sufficient to keep the ecosystem in the desired state, away from its stable equilibrium state (forest). And, the longer this persists, and the better the suite of management techniques becomes, the less the effort needed to achieve this.

The people of the region were able to transform their governance and their scale of operation from the scale of individual farms to the region. The transformed social domain keeps the ecological domain in a quasi-stable flooded-meadow state. The Goulburn-Broken has transformed its social domain, but it cannot prevent the ecological domain from reaching its (unwanted) equilibrium state without also transforming the way the ecosystem is used.

But what if the EKV had wanted something other than the wet meadows? Say they had agreed that they wanted their system to exist as a forest. Had that been the case then a different set of management actions would have been needed but, regardless, the social capital that has been developed in the region would have ensured that their chances of creating and maintaining a desired state are maximized.

Resilience and the KV

This case study demonstrates how the empowerment of locals and, at the same time, the development of governance at a larger scale can contribute to a region's resilience. Looking after what was once an impenetrable swamp has become the glue that binds the local community. The history of the Kristianstad region demonstrates how a traditional piecemeal method of managing our natural resources can lead to the progressive deterioration of the area's environmental values and future options. In the KV, however, this crisis was turned around by the formation of a local body that gave people in the region a voice on how the area was managed.

The adaptive capacity of a social-ecological system is enhanced when complex issues can be dealt with by a network of loosely connected stakeholders located at different levels of society. Such a dynamic structure allows for flexible coordination and cross-scale responses to solving problems because there is experimentation and learning going on across the network. Such experimentation, combined with the networking of knowledge, creates a diversity of experience and ideas for solving new problems. It stimulates innovation and contributes to creating feedback loops at different scales.

6

Creating Space in a
Shrinking World:

Resilience and Sustainability

T he world is shrinking. You can see it all around you. There are
ever increasing numbers of people living off a declining resource
base (Palmer et al. 2004).

- In the Everglades in Florida there are growing problems with
  water quality and a loss of natural habitat. A significant and
  growing chunk of the world-renowned Everglades National
  Park has flipped into a new regime dominated by cattails.
- In the Goulburn-Broken Catchment in Australia, past land management has allowed saline groundwater to rise to just beneath
  the surface of the region's most productive agricultural zone.
  The next sustained wet period will bring it to the surface.
- The coral reefs of the Caribbean are in rapid decline. As they
  are lost, so too is the economic prosperity of the region.
- The Northern Highlands Lake District in Wisconsin is slowly
  losing its natural amenity as it fills up with people.
- The Kristianstad region in Sweden has experienced a progressive decline in water quality and wildlife habitat, and
  needs to maintain a constant effort to ensure that things
  don't slip further.

These are five very different regions experiencing very different
problems but there is a common thread: the population in each region
is growing while the natural diversity, the services provided by ecosystems and a raft of other natural, economic, and social values, are in

decline. It's a thread that entwines our world and one that has us increasingly asking what it means to be sustainable.

What is the key to sustainability? With so many theories and so much discussion on what sustainable development is and how to achieve it, what's special about a resilience framework?

A resilience approach opposes the preoccupation with increased production/yields/returns through increased efficiency (as defined in chapter 1) and control of natural variation. In contrast, resilience thinking captures, and in fact embraces, the dynamic nature of the world. It recognizes the perils of optimizing for particular products or states of a system, and explains why current approaches to managing resources are failing (Holling and Meffe 1996). It describes why and how regions, economies, and ecosystems can change from adaptive, functional systems that provide us with valued goods and services to maladaptive, dysfunctional systems that are more of a liability than an asset to human welfare. It leads to an understanding of critical thresholds in the systems we depend upon, and, once a system has crossed such a threshold into an undesirable regime, resilience thinking explains why it can be so difficult to move out of that condition, and what might be done about it. A resilience approach is about weighing up options, keeping options open, and creating new options when old ones close. In today's world, that's more important than ever.

Declining Options in a Shrinking World

The world is shrinking in many ways. In the introductory chapter we listed some of the global statistics on our growing population and declining resources. But this shrinkage is also about a loss of diversity. As the remorseless process of globalization brings us closer together, joining our cultures, our markets, and our biota, we are becoming increasingly connected and homogenized. The differences that once separated and defined us are getting smaller. Diversity is declining.

Diversity is also declining through direct loss of species, along with the loss of the genes they contain and ecosystems they make up. They are being unwittingly sacrificed as habitat is lost to production and urbanization, and inadvertently lost through the multiple onslaughts of pest invasion, disease, and overharvesting. In 2004, over fifteen thousand species of animals and plants were threatened with extinction (see the IUCN Red List, at www.redlist.org). To give that somewhat dry number some flavor, it includes one in eight species of birds, nearly a quarter of all mammals, a third of amphibians, almost half of turtles and tortoises, and a quarter of conifers. So little is known about what invertebrate and microbe species are being lost that no global estimate exists for them. However, taxonomists believe we're losing many more groups of these organisms than groups of higher animals and plants.

Combine more people and fewer resources with a homogenizing globe that is rapidly losing biodiversity and you end up with shrinking options. There are fewer places to go and less diversity with which to respond to new challenges and unforeseen crises.

Many of the world's leaders and technocrats say this is not a problem. The key to sustainability, they say, is through being more efficient; extract more from less, employ our technological mastery to bridge the growing gap between our needs and available supplies—optimize our way out of the corner we've painted ourselves into. Some conservationists also suggest this is the way forward.

We agree that technical advances and cutting out genuine waste can make an important contribution, but if there is one lesson to be taken from this book it is this: optimization (in the sense of maximizing efficiency through tight control) is a large part of the problem, not the solution. There is no such thing as an optimal state of a dynamic system. The systems in which we live are always shifting, always changing, and in so doing they maintain their resilience—their ability to withstand shocks and to keep delivering what we want.

When we aim to increase the efficiency of returns from some part of the system by trying to tightly control it, we usually do so at the cost of the system's resilience. Other parts of the system begin to change in response to this new state of affairs—a part of the system, now constant, that used to vary in concert with others. A system with little resilience is vulnerable to being shifted over a threshold into a new regime of function and structure. And, as we've seen, this new regime is frequently one that doesn't provide us with the goods and services we want. And, very importantly, it is not a space from which we can easily return.

Short-term optimized harvesting of selected individual stocks has led to the crash of many of the world's most valuable fisheries, and they have not come back. Optimizing agricultural catchments for irrigated agricultural production alone has often led to them becoming degraded, usually forever. Optimizing businesses often leads to an inability to respond to new situations, with costly consequences.

While optimization may increase your returns over short timeframes, it can leave you more and more vulnerable to the changing world around you. It shuts down options.

Creating Space

As we have observed several times, our current approach to dealing with the contraction of safe options is to increase our efficiency, increase our control over change, reduce our waste, and so optimize the systems we manage. Rather than creating more options—more space—this approach only exacerbates the problem. A resilience framework is all about creating space.

There are many pathways into the future for the regions presented as case studies in this book, but those based on being more efficient or extracting greater profit are constraining the future options of the people living in them. On the other hand, pathways that foster experimentation

---

BOX 9 Declining Diversity—Let's Talk Turkey

One of the ways in which our world is shrinking is a loss of biodiversity— the diversity of life, its variety of species, the genes they contain, and the ecosystems they make up. As we lose biodiversity, we lose options.

Unfortunately, because most people in the developed world live in cities far removed from anything like a natural ecosystem, the connection between future options and the loss of a butterfly species in a rainforest on the other side of the globe is bit far removed and tenuous (another way of saying there are loose feedbacks). Maybe this is one of the reasons there isn't a greater effort to protect biodiversity—we simply don't connect it with our own future well-being.

Given this, let us consider an aspect of life's diversity that is a bit closer to home for most of us: domesticated animals and plants that provide us with food. Genetic diversity in agriculture is important. A lack of diversity can increase social vulnerability to pathogens and increase risks to our food supplies (Heal et al. 2004).

In our drive for efficiency, humans are constantly developing domestic livestock breeds and food crops that maximize production of food (and commercial return). And, as we come up with bigger, faster-growing, easier-to-control species, we concentrate our efforts on just these breeds while allowing less-efficient models to die out. In terms of the adaptive

---

cycle (see chapter 4), this is classic K phase behavior in which we are increasingly locked into being more efficient with fewer strategies.

But while these fine-tuned, high-tech breeds produce a lot, they also usually require intensive management and expensive inputs, like high protein feed, medication, and climate-controlled housing. As these high-tech breeds displace the indigenous local breeds that have sustained agriculture for thousands of years, we lose our pool of species and genes that provide us with future options to meet changing circumstances. We're literally putting all our eggs in one basket—becoming totally dependent on a very narrow genetic base that has been selected solely for the goal of maximizing production in a limited set of conditions.

In North America, for example, the commercial white turkey being mass produced on factory farms has been selected for such a meaty breast that it is no longer able to breed on its own. This breed of turkey accounts for 99 percent of all turkeys in the United States today, but it would become extinct in one generation without artificial insemination. And the narrow genetic base makes this breed extremely vulnerable to disturbances such as disease, climate change, and economic shocks.

Increasing genetic uniformity goes hand in hand with the consolidation, control, and ownership of industrial breeding stock. Just three corporate breeders supply the entire world's turkey market. When you mix this level of genetic uniformity with this concentration in ownership it is easy to see how a major part of the world food supply is quite vulnerable.

This process of optimization is happening with livestock breeds everywhere. The diversity of domestic livestock breeds is crashing at an annual rate of 5 percent. The UN's Food and Agriculture Organization (FAO) is describing the process as a threat to international food security because we are losing valuable traits that may prove extremely important in the future of animal breeding.

A few examples of species now under threat of extinction among the many classified by the FAO: the sheep off the Orkney Islands in northern Scotland, which survive exclusively on a diet of seaweed; a cattle breed of Yakut, northern Siberia, which can withstand extreme fluctuations of temperature with very little management; Okulska sheep from southern Poland, which are exceptionally prolific and sometimes produce litters of five or six lambs.

And our loss of diversity is not restricted to animals. The world's wheat supply is dependent on a race between wheat breeders and diseases of wheat, and as more and more wheat varieties disappear the sources of new genes to combat the diseases diminishes. The same applies to rice. And as the diversity of our domestic animals and plants declines and is lost, so too are options for our future.

and innovation maintain the kinds of diversity that build resilience and enhance the social networks operating in a region. These pathways have the greatest chances for achieving long-term well-being. Such approaches create space.

In the Everglades, creating space requires releasing the legislative gridlock to enable a partial reestablishment of natural flood regimes for critical processes, and a focus on ways to minimize phosphorus inflows.

In the Goulburn-Broken Catchment, the time has passed for relying on adaptive capacity. The social-ecological system has crossed a hydrological threshold into a new regime, and creating space in this situation requires transformation. The people of the Goulburn-Broken need to reinvent their social-ecological system. They need to find new ways of making a living.

Creating space in the Caribbean involves the restitution of several functional groups of organisms to allow the coral reefs to regenerate. Unfortunately, the social systems of the Caribbean, being largely poor, far-flung, and poorly connected, don't have the resources or the coordination to mount an effective program that would address the multiple assaults on their precious reef systems. The Caribbean as a region lacks adaptability and is increasingly locked in an undesirable regime.

Scenarios of possible futures have been developed and are being explored by a range of local stakeholders in the Northern Highlands Lake District (NHLD). The exercise is building social capital and enhancing adaptability while avoiding the need for transformation. It is creating space by building networks, enhancing diversity and changing the trajectory it was on.

The Kristianstads Vattenrike did not require the scenario process of the NHLD because the people knew what they wanted, and what they needed—higher adaptability of the social domain; and this was achieved through leadership and the creation of social networks. The result was a transformation in management scale from individual farms to the region, allowing new forms of experimentation.

By focusing on the resilience of a social-ecological system you create space for safe changes in the ecosystem because the system can absorb more shocks and disturbances without crossing a threshold into a new regime. A resilient system has the capacity to change as the world changes while still maintaining its functionality. Resilient systems are forgiving of management mistakes and miscalculations.

BOX 10 Future Directions in Resilience

We don't have all the answers when it comes to understanding resilience and there is much to learn. One new area of investigation, for example, is how to understand and deal with multiple thresholds that occur at different scales in the ecological and social domains of linked systems, and which give rise to cascading regime shifts.

In this book we've demonstrated that a resilience-based approach to using, living in, and governing social-ecological systems is a far sounder basis for long-term human well-being than current approaches to natural resource management. This has been done with examples from local to regional scales—it's at these scales where the research has been done and evidence obtained. The principles, however, extend to national and global scales (Tainter 1998; Diamond 2005).

Some believe the world is poised for a big back loop. Some think we may already be in one (Holling 2004), and that it is expressed at regional scales. Global and continental scale back loops have been experienced in the past. The Black Death and Industrial Revolution were two such events—one resulted in little cheer for humanity, the other in considerable advance. Whatever the future might hold, resilience thinking offers a basis for navigating a graceful passage through the release and reorganization phases that will follow.

What Might a Resilient World Be Like?

We are a long way from the understanding needed to ensure a resilient world. And, as we hope we have made clear, a resilient approach does not sit easily with being prescriptive. Nevertheless, with these two caveats we offer some visions for what a resilient world might be like. (Others have offered comparable lists, e.g., Levin [1999] proposed "eight commandments for environmental management," and we make no claim that ours is unique.)

A resilient world would value:

1. Diversity

A resilient world would promote and sustain diversity in all forms (biological, landscape, social, and economic).

Diversity is a major source of future options and a system's capacity to respond to change and disturbance in different ways (recall response diversity, in particular). Resilient social-ecological systems would celebrate and encourage diversity—offsetting and complementing the existing trend toward homogenizing the world. It would encourage forms of multiple land and other resource use.

2. Ecological Variability

A resilient world would embrace and work with ecological variability (rather than attempting to control and reduce it).

Many of the biggest environmental problems we now face are a result of past efforts to dampen and control ecological variability. Controlling flood levels and preventing species population "outbreaks" are examples embedded in the case studies we have described. Resilience is only maintained by probing its boundaries. A forest that is never allowed to burn soon loses its fire-resistant species, and becomes very vulnerable to a fire.

3. Modularity

A resilient world would consist of modular components.

In resilient systems everything is not necessarily connected to everything else. Overconnected systems are susceptible to shocks and they are rapidly transmitted through the system. A resilient system opposes such a trend; it would maintain or create a degree of modularity.

4. Acknowledging Slow Variables

A resilient world would have a policy focus on "slow," controlling variables associated with thresholds.

By focusing on the key slow variables that configure a social-ecological system, and the thresholds that lie along them, we have a greater capacity to manage the resilience of a system. In so doing it's possible to increase the space (size) of the desirable regime so that the system can absorb more disturbances that might be created by our actions, and so avoid a shift into an undesirable regime. (Alternatively, if we are already in an undesirable regime it enhances our ability to shift out of it.)

5. Tight Feedbacks

A resilient world would possess tight feedbacks (but not too tight).

A resilient social-ecological system would strive to maintain, or tighten, the strength of feedbacks. They allow us to detect thresholds before we cross them. Globalization is leading to delayed feedbacks that were once tighter; the people of the developed world receive weak feedback signals about the consequences of their consumption of developing world products. Feedbacks are loosening at all scales (including the within-scale systems in some of the case studies presented in this book).

6. Social Capital

A resilient world would promote trust, well-developed social networks, and leadership (adaptability).

Resilience in social-ecological systems is very strongly connected to the capacity of the people in that system to respond, together and effectively, to change any disturbance. Trust, strong networks, and leadership are all important factors in making sure this can happen. So, too, is the existence of an institution that has strong penalties for cheaters (Ostrom 1999).

Individually these attributes contribute to what is generally termed "social capital," but they need to act in concert to effect adaptability. As graphically described in Jared Diamond's account of the demise of Greenland's early European settlers (Diamond 2005), strong social ties on their own can be counterproductive, preventing adaptive change. Which leads to the next point.

7. Innovation

A resilient world would place an emphasis on learning, experimentation, locally developed rules, and embracing change.

A resilience approach fosters and encourages novelty and innovation. Our current way of doing things is more about getting better in a decreasing range of activities. Indeed, the current system is mostly about providing subsidies *not* to change, rather than assistance *to* change. Drought assistance and flood relief obviously have a humanitarian component but if they merely perpetuate doing things in the same way they are working against adaptability. A resilient system would subsidize experimentation— trying things in different ways—and offer help to those who are willing to change. Enabling innovation is an important way of creating space.

Resilience thinking is about embracing change and disturbance rather than denying or constraining it. When a back loop begins breaking rigid connections and behaviors, new opportunities open up and new resources are made available for growth. A resilient system is open to this whereas our existing approach is more likely to close down those opportunities. For example, the warning bell to a resilience thinker is increasing preoccupation with process (company policies, public liability, compliance, tort laws, etc.; see "The Dangers of the Late K Phase" in chapter 4, page 85). As systems move inexorably into the late K phase, a resilience approach would advocate initiating a "disturbance," or freeing things up to recapture the dynamics of a growth phase.

8. Overlap in Governance

A resilient world would have institutions that include "redundancy" in their governance structures and a mix of common and private property with overlapping access rights.

Resilient social-ecological systems have many overlapping ways of responding to a changing world. Redundancy in institutions increases the response diversity and flexibility of a system (Ostrom 1999). Such an institutional arrangement fosters a strong awareness and response to cross-scale influences. Totally top-down governance structures with no redundancy in roles may be efficient (in the short term), but they tend to fail when the circumstances under which they were developed suddenly change. More "messy" structures perform better during such times of change.

Access and property rights lie at the heart of many resource-use tragedies. Overlapping rights and a mix of common and private property rights can enhance the resilience of linked social-ecological systems (Dietz et al. 2003).

9. Ecosystem Services

A resilient world would include all the unpriced ecosystem services in development proposals and assessments.

Many of the benefits society gets from ecosystems are either unrecognized or considered "free" (e.g., pollination, water purification, nutrient cycling, and the many others identified by the Millennium Ecosystem Assessment, see www.millenniumassessment.org). These services are often the ones that change in a regime shift and are only recognized and appreciated when they are lost. They are ignored in purely market-driven economies (which, therefore, are inefficient, according to economists' own definition of market efficiency).

Resilience versus Greed

In our opening chapter we broadly discussed the root causes behind the growing list of resource failures around the world. In some situations people have no choice but to allow their resource base to decline (poverty and survival); in others it's allowed to occur willfully (perverse incentives, greed, and corruption); and the third driver of unsustainable development lies in the application of inappropriate models of how the world works (seeking solutions in greater efficiency and a belief in optimal sustainable yields).

This book has specifically addressed the inadequacy of efficiency-driven, optimal-sustainable-yield approaches and puts forward an alternative model based on resilience.

A resilience framework doesn't directly address the problems of greed and the willful consumption of resources. However, a world that embraces the resilience-based themes that we have just outlined goes a long way toward countering many of the problems associated with greed and corruption.

A resilience approach explicitly involves identifying the secondary effects of direct actions, like harvesting. In so doing it highlights the values of otherwise unrecognized ecosystem services (water purification, flood control, pest control, pollination, etc.), making it harder for exploiters to hide the consequences of their greed. In a similar way, because a resilient world sustains diversity and keeps future options open, it makes it harder to justify the conversion of all diversity into single-option solutions that ignore the value of diversity. Greed often leads to a denial of ecological variability in order to profit from short-term development. A resilient world embraces ecological variability rather than attempting to control it and would resist such developments.

A resilient world possesses tight feedbacks, invests in its social capital, and possesses overlapping levels of governance. Cheaters have less space in which to operate and will find that greedy behavior is quickly penalized, often on multiple levels.

And this discussion also relates to the first driver of resource decline—poverty traps. Many of the world's worst examples of poverty traps are, if not due to, at least exacerbated by the other two drivers (greed and inappropriate models). The plight of the Sahelian nomads in Africa was made worse by well-meaning aid agencies that did not understand the complexity of that event-driven pastoral system (Walker and Sinclair 1990). The literature is full of examples of willful conversion and consumption by greedy outsiders of poor peoples' resources. A resilience-based approach would not only deal with the drivers of resource decline associated with the "wrong-model" view of the world, it would also help in resolving the difficult problems associated with drivers of greed and poverty.

We are aware that this will be a gradual process involving a shift in societal ethics (which itself will involve tipping point changes). But it is not a naive outlook. Many people recognize the need. Many already think this way. A framework like resilience thinking will only increase the likelihood of such a shift.

Resilience Thinking

In our opening chapter we observed that there were many pathways into resilience thinking and suggested that readers not worry too much if the finer details of a resilience framework are a bit obscure. We emphasized that what is of more importance is an appreciation of the broader themes that underpin such a framework. Those broader themes revolve around humans existing within linked social and ecological systems. These are complex adaptive systems, and attempts to control or optimize parts of such systems without consideration of the responses that this creates in the broader system are fraught with risk. Much of this book has been spent on attempting to explore the consequences of such an approach.

In the broadest sense, optimizing and controlling components of a system in isolation of the broader system results in a decline in resilience, a reduction in options, and the shrinkage of the space in which we can safely operate. Resilience thinking moves us the other way.

It is our hope that readers who are persuaded of this basic premise will be encouraged to explore the inevitable consequences of such thinking. Even if you are not completely clear on basins of attractions, thresholds, and adaptive cycles, if the concepts of ecological resilience and dynamic social-ecological systems have any resonance then you are in a better position to appreciate what is happening to the world around you.

We began this book by asking: "Why is it that, despite our best intentions, some of the world's most productive landscapes and many of our best-loved ecosystems are in trouble?" The question expanded through the course of the book to include social systems, business enterprises, and social-ecological systems. We hope we have provided some insights to that question.

Resilience thinking is not a panacea for all of the world's problems. It does, however, provide a foundation for achieving sustainable patterns of resource use. The thinking it encapsulates is significantly different from the ruling paradigm of maximizing returns via controlled optimal states in resource management. It encourages us to ask a different set of questions about the way we manage our resources, and therefore ourselves. It challenges many of the assumptions on which we base our actions.

Resilience thinking is a work in progress. To be successful it needs to emerge through people working with their local systems. In this last chapter we have outlined nine attributes of a resilient social-ecological system. The list is not complete. We were tempted to round it off to ten but desisted as we thought it would be more engaging for readers to complete the list for themselves.

So, we finish this book with an invitation. Based on your understanding of what makes a system resilient and sustainable, what is the tenth attribute you would append to our list? We would, seriously, like to know what you think; please send your suggestion to Brian Walker, Brian.Walker@csiro.au.

Postscript for a Resilient World

Over the year that we have been writing *Resilience Thinking* the world has experienced a series of catastrophic natural disasters. In September 2004, four hurricanes of historic proportions hit the Caribbean and Gulf states of the United States (which we alluded to in the case study on the coral reefs of the Caribbean). In December 2004, many countries around the Indian Ocean were devastated by an unprecedented tsunami. Hundreds of thousands of people lost their lives. At the end of August 2005, Hurricane Katrina tore the heart and soul out of New Orleans and the surrounding region; Hurricane Rita followed in September and Wilma in October (during a record season for hurricanes).

There is not space here to dissect the full nature of these disturbances (the full consequences of which may not be appreciated for many years), however it is appropriate in a book on resilience of social-ecological systems to make a brief comment on what we can learn from such events.

Following the tsunami and Hurricane Katrina it was noted by many observers that it was the removal of many protective natural ecosystems (such as mangroves, swamps, barrier silt islands, and coral reefs) that directly contributed to the level of destruction caused by both events. These ecosystems were either replaced by residential, tourist, or commercial development, or destroyed through inappropriate modification of water flows and sedimentation processes. Had these natural buffers been in place the loss of life and property would have been significantly less than what actually occurred.

If the local communities had known more about the ecological drivers of the regions in which they lived, had they embraced the processes of natural change rather than attempted to control them for short-term returns, had they been empowered to make their own decisions about what type of developments were appropriate for their area, and had they been given responsibility to learn to adapt to the changes happening around them then it's likely they would have been much better prepared for the disturbances that hit them.

Of course, that's a lot of "ifs" but each of these aspects contribute to the resilience of these social-ecological systems and each is an acknowledgement that we live in a complex world.

In the aftermath of these events it will be interesting to observe how these social-ecological systems evolve. It's likely that components of these systems will never be the same (as they will have crossed thresholds into new regimes). It's also likely that while there will be opportunities and desires for meaningful restructuring of communities and the ways they are governed, there will also be resistance to change from higher scales.

As we write there is an urgent debate about the probability and consequences of the new (H5N1) Asian bird flu virus changing so as to transmit between humans. If this happens (some say *when*) it's expected a global flu pandemic will result. It raises an important aspect of resilience thinking. Building resilience in a system increases the size of a shock it can experience without a regime shift. But, like the Indonesian territory of Aceh, which took the full force of the tsunami, some shocks will overwhelm whatever resilience a system might have. The effect of Hurricane Katrina on New Orleans would have been less had its resilience been higher, but no mangroves could have interrupted the massive waves that hit Aceh at the end of 2004.

What matters when a system's resilience has been overwhelmed is its adaptive capacity and transformative capacity. From our perspective, that means its ability to reorganize in ways that minimize loss and enhance human well-being. It's an aspect of resilience and sustainability that stood out starkly in the aftermath of Katrina, and is a subject that everywhere deserves much more attention.

Further Reading

Most of the current discussion and analysis on resilience ecology and panarchy are in academic journals and textbooks. While some of these accounts are somewhat technical, if you have any interest in exploring deeper resilience thinking we recommend that you might consider reading some of the books suggested below. To help you in your selection we have annotated these references.

A general reference worth keeping track of is the journal *Ecology and Society* (formerly *Conservation Ecology*), the electronic journal of the Resilience Alliance (see www.ecologyandsociety.org). *Ecology and Society* is a free journal with many papers relevant to this book, which can be downloaded as PDF files.

Also worth investigating is the website of the Resilience Alliance itself (see www.resalliance.org). It contains an extended bibliography and glossary as well as a number of key articles and current activities including a database of examples of thresholds in social-ecological systems.

Adger, W., Hughes, T., Folke, C., Carpenter, S., and Rockstrom, J. 2005. Social-ecological resilience to coastal disasters. *Science* 309:1036–1039.
Presents a timely discussion on how resilient social-ecological systems are inherently better equipped for living with, and learning from, change and unexpected shocks—even when those shocks are on the scale of massive tsunamis and hurricanes.

Berkes, F., Colding, J., and Folke, C., eds. 2003. *Navigating social-ecological systems: Building resilience for complexity and change.* Cambridge University Press, Cambridge, UK.
Drawing on complex systems theory, this book investigates how human societies deal with and build capacity to adapt to change in linked social-ecological systems.

Carpenter, S. R., 2003. *Regime shifts in lake ecosystems.* Ecology Institute, Oldendorf/Luhe, Germany.
Detailed but reasonably accessible textbook on regime shifts, thresholds, and

resilience in lakes. Understanding the dynamics of lake ecosystems provides a powerful example of the value of the resilience framework.

Carpenter, S. R., Walker, B., Anderies, J. M., and Abel, N. 2001. From metaphor to measurement: Resilience of what to what? *Ecosystems* 4:765–781.
This paper compares resilience properties in two contrasting social-ecological systems: lake districts and rangelands. It discusses biodiversity at multiple scales and the existence of institutions that facilitate experimentation, discovery, and innovation.

Folke, C., Carpenter, S. R., Walker, B., Scheffer, M., Elmqvist, T., Gunderson L., Holling, C. S. 2004. Regime shifts, resilience and biodiversity in ecosystem management. *Annual Review in Ecology, Evolution and Systematics* 35:557–581.
Examines the evidence of regime shifts in terrestrial and aquatic environments in relation to resilience of complex adaptive ecosystems and the functional roles of biological diversity in this context.

Gunderson, L. H. and Holling, C. S., eds. 2002. *Panarchy: Understanding transformations in human and natural systems.* Island Press, Washington, D.C.
Over five hundred pages long, this is a series of coauthored chapters that form the basic text of the resilience framework setting out its thinking, context, and consequences. A challenging set of theoretical and empirical perspectives on the dynamic nature of social-ecological systems.

Gunderson, L. H. and Pritchard, L., Jr., eds. 2002. *Resilience and the behavior of large-scale systems. SCOPE Series vol. 60, Island Press, Washington, D.C.*
A collection of case studies of how different kinds of ecosystems have developed various resilience mechanisms to cope with disturbances.

Holling, C. S. 1973. Resilience and stability of ecological systems. *Annual Review of Ecology and Systematics* 4:1–23.
The first scientific paper to define and explain resilience and the stability dynamics of ecosystems.

Holling, C. S. and Meffe, G. 1996. Command and control, and the pathology of natural resource management. *Conservation Biology* 10:328–337.
Explores the consequences of applying increasing levels of top-down, command-and-control management to natural resources, and the implications for resilience.

Holling, C. S. 2004. From complex regions to complex worlds. *Ecology and Society* 9(1):11. Online at www.ecologyandsociety.org/vol9/iss1/art11/.
This paper looks at adaptive cycles operating over a number of scales, and the importance of the back loop in creating space and new opportunities. It's a bit of a personal commentary by C. S. "Buzz" Holling, the scientist who set resilience thinking in motion, on the state of the world and the looming probability of a global release phase. Because it's relatively short (around ten pages long) and readily accessible (you can download it from the Internet), this is a good place to begin following up on this book.

Jen, E., ed. 2005. *Robust design: A repertoire of biological, ecological and engineering case studies*. Santa Fe Institute, Studies in the Science of Complexity, Oxford University Press.
The study of robustness focuses on the ability of a system to maintain specified features when subject to perturbations. Resilience and robustness are related concepts though each has a somewhat different focus and context. This book explores those differences and discusses robustness in complex systems of all shapes and sizes.

Levin, S. 1999. *Fragile dominion*. Perseus Books Group, Cambridge, Massachusetts.
Explores the nature of complex adaptive systems and the connection with evolution and the value of biodiversity.

Millennium Ecosystem Assessment. 2005. *Ecosystems and human well-being*. Island Press, Washington, D.C. Online at www.millenniumassessment.org
This book offers an overview of the Millennium Ecosystems Assessment which focused on how humans have altered ecosystems and changed ecosystem services, and what impact this will have on human well-being. An excellent and, at the time of its writing, up-to-date summary of the state of the world.

Scheffer, M., Carpenter, S., Foley, J. A., Folke, C., and Walker, B. 2001. Catastrophic shifts in ecosystems. *Nature* 413:591–596.
Studies on lakes, coral reefs, oceans, forests, and arid lands show that smooth change can be interrupted by sudden drastic switches to a contrasting state. This paper discusses how a loss of resilience usually paves the way for a switch to an alternate state.

Waldrop M. M. 1992. *Complexity: The emerging science at the edge of order and chaos*. Simon and Schuster, New York.
An account of the development of the science of complexity with a focus on several of the key scientists who pioneered the field and played central roles in the establishment of the Santa Fe Institute. It is an interesting and accessible introduction to the nature and importance of complex adaptive systems.

Walker, B., Carpenter, S., Anderies, A., Abel, N., Cumming, G., Janssen, M., Lebel, L., Norberg, J., Peterson, G. D., and Pritchard, L. 2002. Resilience management in social-ecological systems: A working hypothesis for a participatory approach. *Conservation Ecology* 6(1):14. Online at www.consecol.org/vol6/iss1/art14.
This paper presents an initial attempt to develop a framework and procedure for undertaking an analysis of resilience in a social-ecological system.

Walker, B., Holling, C. S., Carpenter, S., and Kinzig, A. 2004. Resilience, adaptability and transformability in social-ecological systems. *Ecology and Society* 9(2):5. Online at www.ecologyandsociety.org/vol9/iss2/art5/.
A succinct and current description of the logic and terminology of the resilience framework.

References

Adger, W. N., Hughes, T. P., Folke, C., Carpenter, S. R., and Rockstrom, J. 2005. Social-ecological resilience to coastal disasters. *Science* 309: 1036–1039.

Anderies, J. M. 2005. Minimal models and agroecological policy at the regional scale: An application to salinity problems in south-eastern Australia. *Regional Environmental Change* 5:1–17.

Anderies, J. M., Ryan, P., and Walker, B. H. 2005. Loss of resilience, crisis and institutional change—lessons from an intensive agricultural system in south-eastern Australia. *Ecosystems*.

Bellwood, D. R., Hughes, T. P., Folke C., and Nystrom M. 2004. Confronting the coral reef crisis. *Nature* 429:827–833.

Burke, L. and Maidens, J. 2004. *Reefs at risk in the Caribbean*. World Resources Institute. Online at http://marine.wri.org/pubs_description.cfm?PubID =3944.

Carpenter, S. R. 2003. *Regime shifts in lake ecosystems*. Ecology Institute, Oldendorf/Luhe, Germany.

Carpenter, S. R. In press. Seeking adaptive change in Wisconsin's ecosystems. In Waller, D. M. and Rooney, T. P., eds., *The vanishing present: Ecological change in Wisconsin*. University of Wisconsin Press, Madison.

Carpenter, S. R. and Brock, W. A. 2004. Spatial complexity, resilience, and policy diversity: Fishing on lake-rich landscapes. *Ecology and Society* 9(1):8. Online at www.ecologyandsociety.org/vol9/iss1/art8/main.html.

Carpenter, S. R., Levitt, E. A., Peterson, G. D., Bennett, E. M., Beard, T. D., Cardille, J. A., and Cumming, G. S. 2002. *Future of the Lakes*. Illustrations by B. Feeny. Center for Limnology, University of Wisconsin, Madison. Online at http://lakefutures.wisc.edu.

Cumming, D. 1999. Living off biodiversity: Whose land, whose resources and where? *Environment and Development Economics* 4:220–226.

Diamond, J. 2005. *Collapse: How societies choose to fail or succeed.* Viking, New York.

Dietz, T., Ostrom, E., and Stern, P. C. 2003. The struggle to govern the commons. *Science* 303:1907–1911.

Elmqvist, T., Folke, C., Nyström, M., Peterson, G., Bengtsson, J., Walker, B., and Norberg, J. 2003. Response diversity and ecosystem resilience. *Frontiers in Ecology and the Environment* 1(9):488–494.

Folke C., Carpenter S., Walker B., Scheffer M., Elmqvist T., Gunderson L., Holling C. S. 2004. Regime shifts, resilience and biodiversity in ecosystem management. *Annual Review in Ecology, Evolution and Systematics* 35:557–581.

Gardner, T. A., Cote, I. M., Gill, J. A., Grant, A., and Watkinson, A. R. 2003. Long-term region-wide declines in Caribbean corals. *Science* 301:958–960.

Godden, D. 1997. *Agriculture and resource policy.* Oxford University Press, Melbourne, Australia.

Gunderson, L. H. 2001. Managing surprising ecosystems in Southern Florida. *Ecological Economics* 37:371–378.

Gunderson, L. H. and Holling, C. S., eds. 2002. *Panarchy: Understanding trans-formations in human and natural systems.* Island Press, Washington, D.C.

Heal, G., Walker, B., Levin, S., Arrow, K., Dasgupta, P., Daily, G., Ehrlich, P., Maler, K., Kautsky, N., Lubchenco, J., Schneider, S., and Starrett, D. 2004. Genetic diversity and interdependent crop choices in agriculture. *Resource and Energy Economics* 26:175–184.

Hilborn, R. and Waters, C. 1992. *Quantitative fisheries stock assessment: Choice, dynamics and uncertainty.* Chapman and Hall, New York.

Holling, C. S. 1973. Resilience and stability of ecological systems. *Annual Review of Ecology and Systematics* 4:1–23.

Holling, C. S. 1996. Engineering resilience versus ecological resilience. In Schulze, P. C., ed., *Engineering within ecological constraints.* National Academy Press, Washington, D.C.

Holling, C. S. 2004. From complex regions to complex worlds. *Ecology and Society* 9(1):11. Online at www.ecologyandsociety.org/vol9/iss1/art11/.

Holling, C. S. and Meffe, G. 1996. Command and control, and the pathology of natural resource management. *Conservation Biology* 10:328–337.

Levin, S. A. 1998. Ecosystems and the biosphere as complex adaptive systems. *Ecosystems* 1:431–436.

Levin, S. A. 1999. *Fragile dominion.* Perseus Books Group, Cambridge, Massachusetts.

McNeely, J. 1988. *Economics and biological diversity: Developing and using economic incentives to conserve biological resources.* IUCN, Gland, Switzerland.

Olsson, P. and Folke, C. 2001 Local ecological knowledge and institutional dynamics for ecosystem management: A study of Lake Racken watershed, Sweden. *Ecosystems* 4:85–104.

Olsson, P., Folke, C., and Hahn, T. 2004. Social-ecological transformation for ecosystem management: The development of adaptive co-management of a wetland landscape in southern Sweden. *Ecology and Society* 9(4):2. Online at www.ecologyandsociety.org/vol9/iss4/art2/.

Ostrom, E. 1999. Coping with the tragedies of the commons. *Annual Review of Political Science* 2:493–535.

Palmer, M., Bernhardt, E., Chornesky, E., Collins, S., Dobson, A., Duke, C., Gold, B., Jacobson, R., Kingsland, S., Kranz, R., Mappin, M., Martinez, M., Micheli, F., Morse, J., Pace, M., Pascual, M., Palumbi, S., Recihman, O. J., Simons, A., Townsend, A., and Turner, M. 2004. Ecology for a crowded planet. *Science* 304:1251–1252.

Peterson, G. D, Beard, T. D., Jr., Beisner, B. E., Bennett, E. M., Carpenter, S. R., Cumming, G. S., Dent, C. L., and Havlicek, T. D. 2003a. Assessing future ecosystem services: A case study of the Northern Highlands Lake District, Wisconsin. *Conservation Ecology* 7(3):1. Online at www.consecol.org/vol7/iss3/art1/.

Peterson, G. D., Cumming, G. S., and Carpenter, S. R. 2003b. Scenario planning: A tool for conservation in an uncertain world. *Conservation Biology* 17:358–366.

Schumpeter, P. 1950. *Capitalism, socialism and democracy.* Harper and Row, New York.

Scheffer, M., Carpenter, S., Foley, J. A., Folke, C., and Walker, B. 2001. Catastrophic shifts in ecosystems. *Nature* 413:591–596.

Tainter, J. 1988. *The collapse of complex societies* New studies in archaeology, Cambridge University Press, UK.

Walker, B., Holling, C. S., Carpenter, S. R., and Kinzig, A. 2004. Resilience, adaptability and transformability in social-ecological systems. *Ecology and Society* 9(2):5. Online at www.ecologyandsociety.org/vol9/iss2/art5/.

Walker, B. and Meyers, J. A. 2004. Thresholds in ecological and social-ecological systems: A developing database. *Ecology and Society* 9(2):3. Online at www.ecologyandsociety.org/vol9/iss2/art3/.

Walker, B. H. and Sinclair, A. R. E. 1990. Problems of development aid. *Nature* 343:587.

Weatherstone, J. 2003. Lyndfield Park: Looking back, moving forward. Land and Water Australia. Online at www.lwa.gov.au/.

Wood, S., Sebastian, K., and Scherr, S. 2000. *Pilot analysis of global Ecosystems: Agroecosystems.* International Food Policy Research Institute and World Resources Institute, Washington, D.C.

Glossary

Actors: The people who play a role in or have some influence on a social-ecological system. Sometimes referred to as *agents*.

Adaptability: The capacity of actors in a system (people) to manage resilience. This might be to avoid crossing into an undesirable system regime, or to succeed in crossing into a desirable one.

Adaptive cycles: A way of describing the progression of social-ecological systems through various phases of organization and function. Four phases are identified: rapid growth, conservation, release, and reorganization. The manner in which the system behaves is different from one phase to the next with changes in the strength of the system's internal connections, its flexibility, and its resilience.

> **Rapid growth** (r): A phase in which resources are readily available and entrepreneurial agents exploit niches and opportunities.

> **Conservation** (K): Resources become increasingly locked up and the system becomes progressively less flexible and responsive to disturbance.

> **Release** (omega): A disturbance causes a chaotic unraveling and release of resources.

> **Reorganization** (alpha): A phase in which new actors (species, groups) and new ideas can take hold. It generally leads into another r phase.

> The new r phase may be very similar to the previous r phase, or may be fundamentally different. The r to K transition is referred to as the **fore loop**, and the release and reorganization phases are referred to as the **back loop**. Though most systems commonly move through this sequence of the phases, there are other possible transitions.

Basin of attraction: An attractor is a stable state of a system, an equilibrium state that does not change unless it is disturbed. The basin of attraction is all the stable states of the system that tend to change toward the attractor. Figures 3 and 4 show basins of attraction.

Diversity: The different kinds of components that make up a system. In respect to resilience there are two types of diversity that are particularly important.

Functional diversity: Refers to the range of functional groups that a system depends on. For an ecological system this might include groups of different kinds of species like trees, grasses, deer, wolves, and soil. Functional diversity underpins the performance of a system.

Response diversity: Is the range of different response types existing within a functional group. Resilience is enhanced by increased response diversity within a functional group.

Drivers: External forces or conditions that cause a system to change.

Ecosystem services: The combined actions of the species in an ecosystem that perform functions of value to society (pollination, water purification, flood control, etc.).

Equilibrium: A steady-state condition of a dynamic system where the interactions among all the variables (e.g., species) are such that all the forces are in balance, and no variables are changing.

Eutrophication: The enrichment of water by nutrients causing an accelerated growth of algae and other plant life.

Feedbacks: The secondary effects of a direct effect of one variable on another, they cause a change in the magnitude of that effect. A positive feedback enhances the effect; a negative feedback dampens it.

Network: The set of connections (number and pattern) between all the actors in a system.

Panarchy: The hierarchical set of adaptive cycles at different scales in a social-ecological system, and their cross-scale effects (i.e., the effects of the state of the system at one scale on the states of the system at other scales). This nesting of adaptive cycles—from small to large—and the influences across scales is referred to as a panarchy. (See page 90, "Connecting Across Scales.")

Regime: A set of states that a system can exist in and still behave in the same way—still have the same *identity* (basic structure and function). Using the metaphor of the ball in a basin, a regime can be thought of as a system's basin of attraction. Most social-ecological systems have more than one regime in which they can exist.

Regime shifts: When a social-ecological system crosses a threshold into an alternate regime of that system.

Resilience: The amount of change a system can undergo (its capacity to absorb disturbance) and remain within the same regime—essentially retaining the same function, structure, and feedbacks.

Social-ecological systems: Linked systems of people and nature.

State of a system: The state of a system is defined by the values of the *state variables* that constitute the system. For example, if a rangeland system is defined by the amounts of grass, shrubs, and livestock, then the state space is the three-dimensional space of all possible combinations of the amounts of these three variables. The dynamics of the system are reflected as its movement through this space.

Sustainability: The likelihood an existing system of resource use will persist indefinitely without a decline in the resource base or in the social welfare it delivers.

System: The set of state variables (see *State of a system*) together with the interactions between them, and the processes and mechanisms that govern these interactions.

Transformability: The capacity to create a fundamentally new system (including new state variables, excluding one or more existing state variables, and usually operating at different scales) when ecological, economic, and/or social conditions make the existing system untenable.

Thresholds: Levels in underlying controlling variables of a system in which feedbacks to the rest of the system change.

Variables

Controlling variables: Variables in a system (such as nutrient levels in a lake, depth of the water table) that determine the levels of other variables (like algal density or soil fertility).

Fast and slow variables: Controlling ecological variables often tend to change slowly (sediment concentrations, population age structures), while controlling social variables may be fast (e.g., fads) or slow (culture). Slow variables determine the dynamics of the fast variables that are of direct interest to managers. The fast biophysical variables are those on which human use of systems is based, and the fast social variables are those involved in current management decisions or policies.

State variables: See *State of a system*.

About the Authors

BRIAN WALKER is currently program director of the Resilience Alliance and a scientist in Australia's CSIRO Division of Sustainable Ecosystems. He is an ecologist with a particular interest in the resilience of tropical savannas and rangelands. He was born and raised in Zimbabwe, obtained a Ph.D. in Sakatchewan, Canada, lectured at the University of Zimbabwe, and was professor of Botany and director of the Centre for Resource Ecology at the University of Witwatersrand, Johannesburg in South Africa. He moved to Australia in 1985 as Chief of the CSIRO Division of Wildlife and Ecology and stood aside in 1999 to take on his present roles. Brian is a past Chair of the Board of the Beijer International Institute for Ecological Economics in the Swedish Academy of Science, was leader of the International Decade of the Tropics Program on Responses of Savannas to Stress and Disturbance from 1984 to 1990, and led the Global Change and Terrestrial Ecosystems Project of the IGBP from 1989 to 1998. He was awarded the Ecological Society of Australia's Gold Medal in 1999. He has coauthored one book, edited six, written over 150 scientific papers, and is on the editorial boards of five international journals.

DAVID SALT has been writing about science, scientists, and the environment for much of the last two decades. Working with CSIRO Education he developed a newsletter for students into a nationally acclaimed magazine called *The Helix*. He served as Communication Manager for the CSIRO Division of Wildlife and Ecology before becoming the inaugural editor of *Newton* magazine, Australian Geographic's magazine of popular science. He is currently based in Canberra as a freelance science writer where he communicates across the whole spectrum of science. Trained in marine ecology, David is active in environmental education and conservation and has authored a textbook on farm forestry and biodiversity.

Index

Island Press
Board of Directors